T0205379

Representation Learning

Nada Lavrač • Vid Podpečan
Marko Robnik-Šikonja

Representation Learning

Propositionalization and Embeddings

Nada Lavrač
Department of Knowledge Technologies
Jožef Stefan Institute
Ljubljana, Slovenia

School of Engineering and Management
University of Nova Gorica
Vipava, Slovenia

Marko Robnik-Šikonja
Faculty of Computer and Information
Science
University of Ljubljana
Ljubljana, Slovenia

Vid Podpečan
Department of Knowledge Technologies
Jožef Stefan Institute
Ljubljana, Slovenia

ISBN 978-3-030-68819-6   ISBN 978-3-030-68817-2   (eBook)
https://doi.org/10.1007/978-3-030-68817-2

This Springer imprint is published by the registered company Springer Nature Switzerland AG
The registered company address is: Gewerbestrasse 11, 6330 Cham, Switzerland

Dedicated to the One I Love[1]

---

[1] The reader is suggested to search for the song by The Mamas & The Papas on YouTube.

Foreword

Learning a data representation that enables efficient extraction of patterns and learning of predictive models is a crucial first step in applications of machine learning to practical problems of data analysis. Machine learning algorithms often require training data in a tabular format, where each data instance is represented by a fixed-length vector of feature values. To learn from complex data stored in multiple tables of a relational database, corpora of text documents, or a network of interacting data and knowledge nodes, the data analyst has to construct or extract a set of relevant data-representation features, which can express various aspects of the variance in the input data. Thus, the choice of data representation has a profound impact on machine learning performance on a given problem. We find the selection of an appropriate data representation to be of paramount importance for machine learning and data science.

Representation learning aims at developing formal methods for constructing relevant data representation features in complex data. This monograph represents a comprehensible account of representation learning, an important, rapid-evolving field of machine learning. The authors' unique combination of expertise allows them to transcend the survey of various approaches and methods into a novel, unifying framework for representation learning. In particular, Nada Lavrač is among the founding researchers of inductive logic programming and relational learning. Within this monograph, she integrates her seminal work on propositionalization and the numerous contributions to learning from heterogeneous data and knowledge as well as semantic data mining in the context of representation learning. Vid Podpečan excels both in research and programming, allowing him to contribute Jupyter notebooks of Python code, illustrating the practical use of the presented methods. Marko Robnik-Šikonja has an extensive record of in-depth machine learning research with a recent focus on learning for natural language processing and, more specifically, developing cross-lingual embeddings for under-resourced languages.

Through the aforementioned proposal of a unifying framework for representation learning, this monograph integrates two separate and yet closely related venues of research. The first comes from the tradition of inductive logic programming and relational learning, focusing on propositionalization of relational and network data. The second venue builds upon embedding various data types, including text documents and graphs, into vector spaces. Both transform the data into a fixed-length vector and construct features. The resulting vector spaces and their properties are different. The former is binary and logic based, whereas the latter is numeric and distance based. Thus, the former is interpretable, and the latter is not and needs additional explanation. The former does not reduce data dimensionality and is sparse, whereas the latter results in dimensionality reduction and a dense vector representation.

We are delighted to see how this unique integration of the two research venues evolves into a monograph that significantly contributes to the existing machine learning literature. The parts of this monograph on mining text-enriched heterogeneous information networks represent important advances upon the research presented in *Mining Heterogeneous Information Networks*, a book authored by Sun and Han in 2012. The presented approaches to machine learning from ontologies and knowledge graphs significantly update the book on semantic data mining by Ławrynowicz, published in 2017. At the same time, the wide coverage of recent research of embeddings for natural language processing in this monograph nicely complements the recent related books by Pilehvar and Camacho-Collados (2020) on embeddings in natural language processing and by Søgaard et al. (2019) on cross-lingual word embeddings, to name a few. In the last two chapters of this monograph, the authors address a yet-uncovered literature niche of combining propositionalization and embeddings, illustrating the benefits of the unifying framework and introducing two advanced representation learning methods that combine propositionalization with embeddings techniques.

The authors' unique stance on representation learning allows them to cover a wide range of complex data types, from relational data tables, through text, to graphs, networks, and ontologies. The wide coverage of methods and data types makes this monograph appropriate reading for scholars and graduate students in computer science, artificial intelligence, and machine learning. The monograph is a rich source of relevant literature—the reader can find abundant references in each chapter and learn how respective key technologies have evolved, expanded, and been integrated.

The authors conclude each book chapter with labs, which allows the readers to run the presented methods and reuse them for their own data representation endeavors. The resulting repository of Jupyter notebooks for Python promotes the book into an excellent introductory reading for data analysis practitioners. Finally, the unifying framework for representation learning lays a solid foundation for exploring and extending the frontiers of research on data representations for machine learning.

Ljubljana, Slovenia   Ljupčo Todorovski

Tokyo, Japan   Hiroshi Motoda

January 2021

Preface

Improved means for data gathering and storing have resulted in an exponential growth of data collections of increased size and complexity. Moreover, human knowledge encoded in taxonomies, ontologies, and knowledge graphs is constantly growing in size and quality. Consequently, data processing for modern machine learning and data mining requires a significant amount of time and resources. It presents a great challenge for data scientists faced with large amounts of data and knowledge sources of different formats and sizes.

This monograph aims to acquaint the reader with an important aspect of machine learning, called representation learning. Currently, most machine learning algorithms require tabular input, for example, decision trees, support vector machines, random forests, boosting, and neural networks. For this reason, the research community has created many automated data transformation techniques applicable to scenarios involving data and background knowledge in heterogeneous formats. The monograph provides an overview of modern data processing techniques, which enable data fusion from different data formats into a tabular data format. In the transformed data representation, data instances are represented as vectors in the symbolic and/or numeric space. The key element of the success of modern data transformation methods is that similarities of original instances and their relations are expressed as distances and directions in the target vector space, which allows machine learning algorithms to group similar instances based on their distance and distribution.

The work establishes a unifying approach to data transformations of various data formats addressed in modern data science, enabling the reader to understand the common underlying principles. The focus is on data transformation variants developed in relational learning and inductive logic programming (named propositionalization) and modern embedding approaches, which are particularly popular in mining texts, networks, ontologies, and knowledge graphs. While both approaches enable the user to perform data transformations from different data formats to a single table representation, their underlying principles are different: embeddings perform dimensionality reduction, resulting in lower-dimensional vectors capturing the semantics of the data, whereas propositionalization flattens first-order logical features, which have higher descriptive power than the attribute values. In summary, both propositionalization and embeddings are data transformation techniques, but result in different tabular outputs:

- *Propositionalization* is used to get symbolic vector representations from relational databases, as well as from a mixture of tabular and networked data, including ontologies.
- *Embeddings* are used to get numeric vector representations from many complex data forms, performing data transformations that preserve the semantic information contained in texts, relations, graphs, and ontologies.

Other approaches to data transformation and fusion, including data transformations into a network format, are not covered in this text.

The audience of this monograph are students as well as machine learning practitioners.

- The first group of readers is students of computer science, engineering, or statistics that wish to understand and master a deluge of data embedding techniques through a unifying approach. The monograph may serve as supplementary reading in courses on data mining, machine learning, data science, and statistical data analysis.
- The second group of readers are data science and machine learning practitioners. The expected level of technical proficiency suits both inexperienced users as well as experienced data scientists. Most practitioners are unaware of the multitude of available data embedding techniques, which we try to present in an accessible manner.

While there are several books addressing related topics on data preprocessing and data transformation, none of them provides a comprehensive overview of transformation techniques for heterogeneous data embeddings in a wider data mining context. After the introduction, the monograph starts with an overview of related machine learning research, followed by the presentation of text embedding approaches. Structured data representations and appropriate mechanisms for complex feature construction are addressed in the next chapters, presenting data transformations needed for mining multi-relational data, graphs, ontologies, and heterogeneous information networks. This is followed by the presentation of approaches unifying propositionalization and embeddings into joint data representation pipelines, and a summary of the unifying and differentiating aspects of propositionalization and embeddings, providing arguments why and how propositionalization and embeddings can be considered as two sides of the same coin in the context of representation learning.

As its distinguishing feature, this monograph covers transformation of heterogeneous data, including tables, relations, text, networks, and ontologies, in a unified data representation framework. The monograph combines intuitive presentations of the main ideas underlying individual representation learning approaches, their technical descriptions and explanations, as well as practical access to code, executable use cases, and examples. The reader will be able to run and reuse illustrative examples provided in the monograph's repository of Jupyter Python notebooks, which have been carefully prepared to be up to date, nontrivial, but easy to understand. Most chapters contain parts of our original research work, written in collaboration with doctoral students and other researchers. Most notably, we acknowledge the contributions of Anže Vavpetič and Filip Železný, with whom we have successfully collaborated in propositionalization and semantic data mining research. Doctoral students and researchers, who have significantly contributed to the research presented in this monograph, include Miha Grčar, Jan Kralj, Matic Perovšek, Blaž Škrlj, and Matej Ulčar. We are grateful also to outstanding machine learning researchers, Ljupčo Todorovski and Hiroshi Motoda, for contributing the inspiring foreword to this monograph.

We wish to acknowledge our institutions, Jožef Stefan Institute, University of Nova Gorica, and University of Ljubljana, Faculty of Computer and Information Science, Slovenia, for providing stimulating working environments. We acknowledge the financial support of the Slovenian Research Agency project Semantic Data Mining for Linked Open Data N2-0078 and the core research programs P2-0103 and P6-0411. We gratefully acknowledge also the funding from the European Union's Horizon 2020 research and innovation program under grant agreement No 825153 (EMBEDDIA). We are also grateful to Springer Nature for the support in publishing this monograph.

Ljubljana, Slovenia   Nada Lavrač

Ljubljana, Slovenia   Vid Podpečan

Ljubljana, Slovenia   Marko Robnik-Šikonja

January 2021

Contents

Chapter 1
Introduction to Representation Learning

Data scientists are faced with large quantities of data in different forms and sizes. Modern data processing techniques enable data fusion from different formats into a tabular data representation, where instances are represented as vectors. This form is expected by standard machine learning techniques such as rule learning, support vector machines, random forests, or deep neural networks. The key element of the success of modern representation learning methods, which transform data instances into a vector space, is that similarities of the original data instances and their relations are expressed as distances and directions in the target vector space, allowing for similar instances to be grouped based on these properties.

This chapter starts by motivating the need for representation learning aimed at transforming input datasets into a tabular format in Sect. 1.1, and by situating representation learning as an automated data transformation step into the overall knowledge discovery process in Sect. 1.2. Two main families of representation learning methods, propositionalization and embeddings, are defined and contextualized in the framework of information representation levels in Sect. 1.3. Next, evaluation of representation learning approaches is presented Sect. 1.4, followed by a brief survey of a variety of representation learning methods in Sect. 1.5. The chapter concludes with the outline of this monograph in Sect. 1.6.

1.1 Motivation

Machine learning is the key enabler for computer systems to progressively improve their performance when helping humans to handle difficult problem-solving tasks. Nevertheless, current machine learning approaches only come half-way in helping humans, as humans still have to formulate the problem and prepare the data in the form best suited for processing by powerful machine learning algorithms.

© Springer Nature Switzerland AG 2021
N. Lavrač et al., *Representation Learning*,
https://doi.org/10.1007/978-3-030-68817-2_1

Most of the best performing machine learning algorithms, like support vector machines (SVMs) or deep neural networks (DNNs), assume numeric data and outperform symbolic approaches in terms of predictive performance, efficiency, and scalability. The dominance of numeric algorithms started in the 1980s with the advent of backpropagation and neural networks (Rumelhart et al. 1986), continued in the late 1990s and early 2000s with SVMs (Cortes and Vapnik 1995), and reached the current peak with deep neural networks (Goodfellow et al. 2016). DNNs are currently considered the most powerful learners for solving many previously unsolvable learning problems in computer vision (face recognition rivals humans' performance), game playing (a program has beaten a human champion in the game of Go), and natural language processing (successful automatic speech recognition and machine translation).

While the most powerful machine learning approaches are numeric, humans perceive and describe real-world problems mostly in symbolic terms, using various data representation formats, such as graphs, relations, texts or electronic health records, mostly involving discrete representations. However, if we are to harness the power of successful numeric deep learning (DL) approaches for discrete learning problems, discrete data should be transformed into a form suitable for numeric learning algorithms.

The viewpoint of addressing real-world problems as numeric has a rationale even for discrete domains, as many symbolic learners perform generalizations based on object similarity. For example, in graphs, nodes can represent similar entities or have connections with similar other nodes; in texts, words can appear with similar contexts or play the same role in sentences; in medicine, patients may have similar symptoms or similar disease histories. Numerous machine learning algorithms use such similarities to generalize and learn, including classical bottom-up learning approaches such as hierarchical clustering, as well as symbolic learners adapted to top-down induction of clustering trees (Blockeel et al. 1998). To exploit the power of modern machine learning algorithms (like SVMs and DNNs) to process the inherently discrete data, one has to transform discrete data into numeric data in a way that allows for similarities between objects to be preserved and expressed as distances in the transformed numeric space.

Contemporary data transformation approaches that prepare numeric tabular data for machine learning algorithms are referred to as *embeddings*. Nevertheless, as demonstrated in this monograph, symbolic data transformations remain highly relevant. The role of *propositionalization*, a symbolic approach to transforming relational instances into feature vectors, is not only to enable contemporary machine learning algorithms to induce better predictive models but also to allow descriptive data mining approaches to discover interesting human-comprehensible patterns in symbolic data.

As this monograph demonstrates, albeit propositionalization and embeddings represent different types of data transformations, they can be viewed as the *two sides of the same coin*. Their main unifying element is that they transform the input data into a tabular format and express the relations between objects in the original space as distances (and directions) in the target vector space.

1.2   Representation Learning in Knowledge Discovery

Representation learning is a set of techniques that allow automatic construction of data representations needed for machine learning (Bengio et al. 2013). This learning task is essential in modern deep learning approaches, where it replaces manual feature engineering.

Representation learning is motivated by the fact that machine learning tasks, such as classification, often require input data of a prescribed data format. However, real-world data such as images, video, and sensor data are not intrinsically available in the form convenient for effective learning. Most learning algorithms expect the input format, where each instance is represented with a vector of features. Therefore, the whole dataset forms a table, with rows corresponding to instances represented by vectors of feature values.

This section introduces representation learning as a separate automated data transformation step in the overall process of knowledge discovery in databases, whose main processing step is data analysis, performed by machine learning and/or data mining algorithms. We start by briefly introducing machine learning and data mining, as well as the process of Knowledge Discovery in Databases (KDD) in Sect. 1.2.1, and focus on the automated data transformation step of the overall KDD process in Sect. 1.2.2.

1.2.1   Machine Learning and Knowledge Discovery

Machine Learning (ML) (Mitchell 1997) is a research area of computer science, mainly concerned with training of classifiers and predictive models from labeled data. On the other hand, *data mining* (Han and Kamber 2001) is a research area in the intersection of computer science and database technology, mainly concerned with extracting interesting patterns from large data stores of mainly unlabeled data. Furthermore, *data science* (Leek 2013) tries to unify statistic, machine learning, and data mining methods to understand real-life phenomena through data analysis.

The terms data mining and machine learning are frequently used interchangeably, given no strict limitations between them. The research community usually uses the term machine learning, while data mining is mostly used in industry and by machine learning practitioners. Whatever term used, both terms refer to a very narrow data analysis step in the overall process of extracting knowledge from data, referred to as *Knowledge Discovery in Databases* (KDD) (Piatetsky-Shapiro and Frawley 1991; Fayyad et al. 1995) or more recently, *Data Science* (DS) (Leek 2013). The entire process deals with the extraction of interesting (nontrivial, implicit, previously unknown, and potentially useful) information from data in large databases (Fayyad et al. 1996).

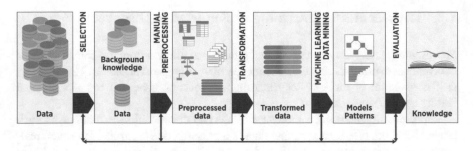

Fig. 1.1 The overall KDD process, composed of several steps: data selection, manual data preprocessing, automated data transformation, machine learning or data mining, and model/pattern evaluation

The overall KDD process of finding and interpreting models or patterns in data involves the repeated application of several steps (Piatetsky-Shapiro and Frawley 1991), illustrated in Fig. 1.1.

Data selection.   This step refers to developing an understanding of the application domain, the relevant prior knowledge, the end user's goals, selecting a dataset focusing on a subset of variables, or data samples, on which discovery is to be performed.

Data preprocessing.   This step refers to removing noise or outliers, collecting the necessary information to model or account for noise, handling missing feature values, accounting for time sequence information and known changes in the data.

Data transformation.   This step refers to automated feature engineering (feature extraction or construction) and the formation of the final training set, to be used as input to a machine learning algorithm. This step is in the focus of this monograph.

Machine learning or data mining.   This step refers to applying the most appropriate machine learning and data mining tools for summarization, classification, regression, association, or clustering, to construct models or find patterns in the data. Note that in this monograph, we use the terms machine learning and data mining interchangeably to denote this crucial step of the overall KDD process.

Evaluation.   This step refers to quality assessment of the results, including patterns or model validation, patterns or model explanation, their visualization, and removal of potentially redundant patterns. While each of the previous steps might be evaluated on its own (e.g., the quality of the representation in the data transformation step), this is the final overall evaluation. From the point-of-view of the previous steps, this evaluation is often referred to as *extrinsic* evaluation on a *downstream* learning task.

Terms machine learning and data mining are distinct from KDD. Machine learning and data mining refer to a single step of the KDD process, i.e. applying the algorithms to extract models or patterns from data. On the other hand, KDD refers to the overall process of discovering useful knowledge in data. It includes the choice of data encoding schemes, data selection and sampling, manual data preprocessing, and automated data transformation (aimed at fusing potentially heterogeneous data and data projection into a joint data representation) before the machine learning/data mining step. It also involves evaluating models/patterns and interpreting the newly constructed knowledge, which follows the machine learning or data mining step.

In this monograph, we consider data preprocessing and data transformation as two separate subprocesses of data preparation.

Data preprocessing. Data preprocessing refers to a step of data cleaning, instance selection, normalization, handling missing attribute values, controlling out-of-range values and impossible attribute-value combinations, or handling noisy or unreliable data, to name just some of the types of data irregularities encountered in processing real-life data. Data preprocessing can be manual, automated, or semi-automated.

Data transformation. Data transformation refers to the automated representation learning step, consisting of feature construction for compact data representation, allowing for data projection aimed at finding invariant representations of the data in the multi-dimensional vector space.

In the next section we focus on the latter, automated data transformation.

1.2.2 Automated Data Transformation

Automated transformation of data, present in hctcrogeneous types and formats, should result in a uniform tabular data representation. We refer to this specific automated data processing task as *data transformation* or *representation learning*. The data transformation task, illustrated in Fig. 1.2, is defined as follows.

Definition 1.1 (Data Transformation) Data transformation refers to the automated representation learning step of the KDD process that automatically transforms the input data and the background knowledge into a uniform tabular representation, where each row represents a data instance, and each column represents one of the automatically constructed features in a multidimensional feature space.

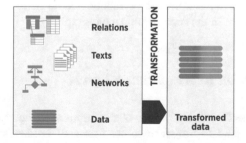

Fig. 1.2 Representation learning can be considered a data transformation task. Here we illustrate the case where the target data used by the learner for model/pattern construction is in tabular form, and the heterogeneous background knowledge includes a relational database, a document corpus and/or a network of interconnected entities (e.g., a knowledge graph)

In the above definition, we decided to distinguish between *data* and *background knowledge*. This is an intentional decision, although it could be argued that in some settings, we could refer to both as data. Let us provide an operational distinction between data and background knowledge.

- *Data* is considered by the learner as the target data from which the learner should learn a model (e.g., a classifier in the case of class labeled data) or a set of descriptive patterns (e.g., a set of association rules in the case of unlabeled data).
- *Background knowledge* is any additional knowledge used by the learner in model or pattern construction from the target data. Simplest forms of background knowledge define hierarchies of features (attribute values), such as color *green* being more general than *light green* or *dark green*. More complex background knowledge refers to any other declarative prior domain knowledge, such as knowledge encoded in relational databases, knowledge graphs, or domain-specific taxonomies and ontologies, such as the Gene Ontology (GO), which included 44 508 GO terms, 7 765 270 annotations, 1 464 358 gene products, and 4593 species in its 2020-05-02 release.

The addressed data transformation setting is applicable in various data science scenarios involving relational data mining, inductive logic programming, text mining, graph, and network mining, as well as tasks requiring data fusion of a variety of data types and formats and their transformation into a joint data representation format.

Representation learning approaches enable and simplify machine learning on many complex data types (texts, relations, graphs, and ontologies). The performed data transformations preserve the semantic similarities contained in data.

1.3   Data Transformations and Information Representation Levels

This section first outlines the three levels of cognitive representations proposed by Gärdenfors (2000), followed by defining data transformations into two of these levels, i.e. data transformations into the symbolic and numeric representation space, respectively, which are in the focus of this monograph.

1.3.1   Information Representation Levels

Since the most powerful ML algorithms take as input numeric representations, users of ML algorithms tend to transform other forms of human knowledge into the numeric representation space. Interestingly, using numeric representations of data for modern ML algorithms also started prevailing in data mining tasks traditionally addressed by learners designed for symbolic data representations, which are most frequently used for encoding the stored human knowledge.

The distinction between the symbolic and numeric representation space mentioned above can be further clarified in terms of the *levels of cognitive representations*, introduced by Gärdenfors (2000), i.e. the neural, spatial and symbolic representation levels. In his theory, Gärdenfors assumes that when modeling cognitive systems in terms of information processing, all three levels are connected: starting from the sensory inputs at the lowest neural representation level, resulting in spatial representations at the middle conceptual spaces level, up to symbolic representations at the level of language.

Neural. This representation level corresponds to the sub-conceptual connectionist level. At this level, information is represented by activation patterns in densely connected networks of primitive units. This enables concepts to be learned from the observed data by modifying the connection weights between the units.

Spatial. This representation levelrepresentation level!spatial is modeled in terms of Gärdenfors' conceptual spaces. At this level, information is represented by points or regions in a conceptual space built upon some dimensions that represent geometrical, topological, or ordinal properties of the observed objects. In spatial representations, the similarity between concepts is represented in terms of the distances between the points or regions in a multidimensional space. Concepts are learned by modeling the similarity of the observed objects.

Symbolic. At this representation representation level!symbolic level, information is represented by the language of symbols (words). The meaning is internal to the representation itself, i.e. symbols have meaning only in terms of other symbols. At the same time, their semantics is grounded at the spatial level. Here, concepts are learned by symbolic generalization rules.

From the perspective of this monograph, the above levels of cognitive representations introduced by Gärdenfors (2000) provide a theoretical ground to separate learning and data transformation approaches into three categories based on the levels of their output representation space: neural, spatial, and symbolic. However, given the scope of this monograph, we do not consider *neural transformations*, and focus only on the other two data transformation types:

Symbolic transformations. Symbolic transformations, which denote data transformations into a symbolic representation space, are in this monograph referred to as *propositionalization*.

Numeric transformations. Numeric transformations, which denote data transformations into a spatial representation space, are in this monograph referred to as *embeddings*.

These two data transformation tasks are briefly introduced in Sects. 1.3.2 and 1.3.3, respectively.

1.3.2   Propositionalization: Learning Symbolic Vector Representations

In symbolic learning, the result of a machine learning or data mining algorithm is a predictive model of a set of patterns described in a symbolic representation language, resulting in symbolic human-understandable models and patterns. Symbolic machine learning approaches include classification rule learning (Michalski et al. 1986; Clark and Niblett 1989), decision tree learning (Quinlan 1986), association rule learning (Agrawal and Srikant 1994), and learning logical representations by relational learning and inductive logic programming algorithms (Muggleton 1992; Lavrač and Džeroski 1994; De Raedt 2008).

To apply a symbolic learner, the data is typically transformed into a single tabular data format, where each row represents a single data instance, and each column represents an attribute or a feature. Such transformation into symbolic vector space (i.e. a symbolic data table format) is well known in the Inductive Logic Programming (ILP) and relational learning communities, where it is referred to as *propositionalization*. The propositionalization task is formally defined below.

Definition 1.2 (Propositionalization)

Given: Input data of a given data type and format, and heterogeneous background knowledge of various data types and formats.

Find: A tabular representation of the data enriched with the background knowledge, where each row represents a single data instance, and each column represents a feature in a d-dimensional *symbolic feature space* F^d.

The distinguishing property of data transformation via propositionalization, compared to embeddings-based data transformations addressed in Sect. 1.3.3, is that propositionalization results in the construction of interpretable symbolic features, as illustrated in Example 1.1.

Example 1.1 (Simple Propositionalization) Suppose we have data about persons (presented in a tabular format, shown in Table 1.1) and about their family relations in the form of a family tree. The tree, which is usually represented as a graph, can also be represented in a tabular form (see Table 1.2). Propositionalization is a data transformation, which takes as input the tabular data from Table 1.1, enriched by background knowledge family tree relations shown in Table 1.2, and transforms it into a tabular representation described by binary features, as shown in Table 1.3.

Table 1.1 Persons described by age, gender and education level, using an attribute-value representation

Person	Age	Gender	Education	Class
Paul	18	male	secondary	No
Alice	42	female	university	Yes
Tom	45	male	professional	Yes

Table 1.2 Relations between persons in tabular form

Person 1	Person 2	Relation
Alice	Paul	hasChild
Tom	Paul	hasChild
Tom	Alice	married

Table 1.3 Propositionalized representation of persons, described with binary features

Person	Age < 20	IsMale	HasUniDegree	HasChild	IsMarried	Class
Paul	1	1	0	0	0	0
Alice	0	0	1	1	1	1
Tom	0	1	0	1	1	1

Let us explain the definition of propositionalization in more detail. The definition implies that propositionalization is characterized by the construction of symbolic (interpretable) features. However, their values in the transformed data tables are of two different kinds.

- Traditionally, propositionalization resulted in the construction of logical (binary) features. In this case, each value in each column of the transformed data table is binary, i.e. either 1 or 0 (or in logical terms, either *true* or *false*), as illustrated in Table 1.3 of Example 1.1 (as another example see e.g., Table 4.4 in Sect. 4.3).
- Note, however, that symbolic features can be weighted by their importance. In this case, each value in each column of the transformed data table is numeric, with values $v \in [0, \infty)$ (as is the case in the middle table of Table 4.7 in Sect. 4.5), or with values $v \in [0, 1]$ (as is the case in the bottom table of Table 4.7 in Sect. 4.5).

In the latter, the output are numeric vectors. However, what distinguishes these numeric vectors are interpretable symbolic features used as variables, as opposed to non-interpertable features used in numeric vectors learned in embeddings (addressed in Sect. 1.3.3).

1.3.3   Embeddings: Learning Numeric Vector Representations

The last 20 years have been witnessing an increasing dominance of statistical machine learning and pattern-recognition methods, including neural network learning (Rumelhart and McClelland 1986), support vector machines (Vapnik 1995; Schölkopf and Smola 2001), random forests (Breiman 2001), boosting (Freund and Schapire 1997), and stacking (Wolpert 1992).

These statistical approaches are quite different from the symbolic approaches mentioned in Sect. 1.3.2. However, many approaches cross these boundaries, including selected decision tree learning (Breiman et al. 1984) and rule learning (Friedman and Fisher 1999) algorithms that are firmly based in statistics. Ensemble

Table 1.4 Dense numeric
representation of Persons
from Table 1.3

Person	F_1	F_2	Class
Paul	0.78510	0.11112	0
Alice	0.34521	0.22314	1
Tom	0.91767	0.23412	1

techniques such as boosting (Freund and Schapire 1997), and bagging (Breiman 2001) also combine the predictions of multiple logical models on a sound statistical basis.

To apply a statistical learner, the data is typically transformed into a single table data format, where each row represents a single data instance. Each column is a numeric attribute or a numeric feature, with some predefined range of numeric values. Such transformation into the numeric vector space (i.e. a numeric data table format) is referred to as *embedding* in the data science community. The embedding task is formally defined below.

Definition 1.3 (Embeddings)

Given: Input data of a given data type and format, and heterogeneous background knowledge of various data types and formats.
Find: A tabular representation of the data enriched with the background knowledge, where each row represents a single data instance, and each column represents one of the dimensions in the d-dimensional *numeric vector space* $\mathbb{R}^d$.

Example 1.2 (Simple Embeddings) Let us reuse the data from Example 1.1 about persons and their family relations. One way to embed this dataset could be by starting from the propositionalized data representation given in Table 1.3 and embedding it into the numeric space, as shown in Table 1.4. This can be achieved e.g., by applying the PropStar algorithm (see Sect. 6.2.1) to transform the propositionalized data representation into a lower-dimensional numeric space, preserving the relational information.

In this monograph, embeddings mostly refer to dense numeric representations involving relatively few non-interpretable numeric features. However, a few exceptions, e.g., one-hot-encoding or bag-of-words representation, introduced in Sect. 3.2.1, are sparse and encode symbolic features, yet can be treated either as propositionalized vectors or as embeddings, depending on the context of their use.

1.4   Evaluation of Propositionalization and Embeddings

Given that representation learning is itself a learning task, i.e. the task of learning an appropriate representation of the data, we can evaluate the success of representation learning in terms of its performance and interpretability, addressed in Sects. 1.4.1 and 1.4.2, respectively.

1.4.1   Performance Evaluation

Performance evaluation of propositionalization and embedding approaches can be evaluated in two ways: through their *extrinsic* and *intrinsic* evaluation.

Extrinsic evaluation.   In the extrinsic evaluation of propositionalization and embeddings, the quality of data transformations is evaluated on downstream learning tasks, such as classifier construction. For example, in the classification setting, different approaches with different parameters are used in the data transformation step of the KDD process. When they are compared for a given classifier, using some quality assessment measure such as classification accuracy or precision (see Sect. 2.5), the highest classification performance of the constructed classifier implicitly denotes the best approach used in the data transformation step.

Intrinsic evaluation.   Intrinsic evaluation of transformations refers to their evaluation on some accessible and computationally manageable task. Regardless of the types of input objects, the most common intrinsic evaluation of propositionalization and embeddings of entities concerns the assessment of whether the *similarities* of the input entities (training examples) described in the original representation space are preserved in terms of the similarities of the transformed representations. Another intrinsic evaluation task measures the ability to reconstruct the original data after a given transformation (evaluated by some loss function using the data reconstruction error). Nevertheless, most frequently, the intrinsic evaluation of transformation approaches depends on the specific transformation approach and on the specific task designed for evaluating the quality of data transformations. For example, for text embeddings, the *word analogy* task can be used for intrinsic evaluation of text embeddings (see Sect. 3.6).

In summary, both evaluation approaches have advantages and disadvantages. While extrinsic evaluation is best aligned with the overall purpose of the learning task, it might be computationally demanding as it requires the execution of the entire KDD process. Moreover, the interactions between different components of the KDD process might prevent the fully justified assessment of the data transformation step itself. On the other hand, given that the purpose of intrinsic evaluation is to get quick information on the quality of produced transformations, there may be unfortunately no guarantee that good performance on an intrinsic task correlates well with the performance on a downstream task of interest.

1.4.2   Interpretability

Concerning the interpretability of results, propositionalization approaches are mostly used with symbolic learners whose results are interpretable, given the interpretability of features used in the transformed data description. For embedding-based methods, given non-interpretable numeric features, specific mechanisms need to be implemented to ensure results explanation (Robnik-Šikonja and Kononenko 2008; Štrumbelj and Kononenko 2014).

A recent well-known approach, which can be used in the final step of the KDD process, is results explanation using the SHAP approach (Lundberg and Lee 2017). In this approach, Shapley values offer insights into instance-level predictions by assigning fair credit to individual features for participation in prediction-explaining interactions. Explanation methods such as SHAP are commonly used to understand and debug black-box models. We refer the reader to the work of Lundberg and Lee (2017) for a detailed overview of the method. We briefly address the interpretability of transformations in Chap. 7, where we discuss the unifying aspects of propositionalization and embeddings.

1.5 Survey of Automated Data Transformation Methods

While there are many algorithms for transforming data into spatial (numeric) representations, it is interesting that recent approaches rely on deep neural networks, thereby harnessing the neural representation level to transform symbolic representations into numeric representation. Below we list the main types of approaches that perform data transformations.

Propositionalization methods.   These methods are used to get tabular data from relational databases as well as from a mixture of tabular data and background knowledge in the form of logic programs or networked data, including ontologies. These transformations were mostly developed within the relational learning and inductive logic programming communities and are still actively researched and used. Propositionalization methods do not perform dimensionality reduction and are most often used with data mining and symbolic machine learning algorithms. We discuss these methods in Chap. 4.

Graph traversal methods.   Many complex datasets can be represented as graphs, where nodes represent data instances and edges represent their relations. Graphs can be homogeneous (consisting of a single type of nodes and relations) or heterogeneous (consisting of different types of nodes and relations). To encode a graph in a tabular form by preserving the information about the relations, various graph encoding techniques were developed, such as propositionalization via random walk graph traversal, representing nodes via their neighborhoods and communities (Plantié and Crampes 2013). These approaches are frequently used for data fusion in mining heterogeneous information networks. We discuss them in Sect. 5.1.

Matrix factorization methods.   When data is not explicitly presented in the form of relations, but the relations between objects are implicit, given by a similarity matrix, the objects can be encoded in a numeric form using matrix factorization. As an example, take Latent Semantic Analysis used in text mining, which factorizes a word similarity matrix to represent words in a vector form, presented in Sect. 3.2.5. Deep neural networks largely superseded these types of embeddings, as they do not preserve the similarities between entities but rather construct a prediction task and forecast similarity.

Neural network-based methods.   In neural networks, the information is represented by activation patterns in interconnected networks of primitive units. This enables that concepts are gradually learned from the observed data by modifying the connection weights between the hierarchically organized units. These weights can be extracted from neural networks and used as a spatial representation that transforms relations between entities into distances. Recently, this approach became a prevalent way to build representation for many different types of entities, e.g., texts, graphs, electronic health records, images, relations, recommendations, etc. Selected neural network-based methods are outlined below.

Text embeddings.   These embeddings use large corpora of documents to extract vector representations for words, sentences, and documents. The first neural word embeddings like word2vec (Mikolov et al. 2013) produced one vector for each word, irrespective of its polysemy (e.g., for a polysemous word like bank, word2vec produces a single representation vector, and ignores the fact that bank can present both a financial institution and a land sloping down to a water mass). The word2vec learning task is to predict words based on their given neighborhood. Recent developments like ELMo and BERT take a context of a sentence into account and produce different word vectors for different contexts of each word. Text embeddings are addressed in Chap. 3.

Graph and heterogeneous information network embeddings.   These embeddings capture the structure of the graph by using convolution filters that capture the most relevant shapes to predict links in the graph. This way, the weights of the neural network are used as numeric vectors. We describe embeddings for several types of graphs, from simple graphs to heterogeneous information networks, in Chap. 5.

Entity embeddings.   These embeddings can use any similarity function between entities to form a prediction task for a neural network. Pairs of entities are used as a training set for the neural network, which forecasts whether two entities are similar or dissimilar. The weights of the trained network are then used in the embeddings. As this is probably the most general approach to embeddings, used in many different tasks, we describe it more comprehensively in Sect. 6.1.

Other embedding methods.   Other forms of embeddings were developed by different communities that observed the need for improved representations of (symbolic) data. For example, Latent Dirichlet Allocation (LDA) (Blei et al. 2003) used in text analysis learns distributions of words for different topics. These distributions can be used as an effective embedding for words, topics, and documents. Feature extraction methods form a rich representation of instances by projecting them into a high dimensional space (Lewis 1992). Another example of (implicit) transformation into high dimensional space is the kernel convolutional approach proposed by Haussler (1999), which introduces the idea that kernels can be used for discrete structures by iteratively applying convolution and kernels to smaller parts of the data structure. Convolutional kernels exist for sets, graphs, trees, strings, logical interpretations, and relations (Cumby and Roth 2003; Gärtner et al. 2004). This allows methods such as SVM or Gaussian Processes to work with relational data. Most of these embeddings are recently superseded or merged with neural networks.

All the above approaches perform data transformations from different data formats to a single table representation. However, their underlying principles are different: while factorization and neural embeddings perform dimensionality reduction, resulting in lower-dimensional feature vector representations capturing the semantics of the data, propositionalization results in a vector representation using relational features with a higher generalization potential than the features used in the original data representation.

1.6 Outline of This Monograph

This monograph serves as an introduction to propositionalization and embedding techniques, while covering also the recent advances. In this chapter, we introduced propositionalization and embeddings as data transformation approaches used in the KDD process (Piatetsky-Shapiro and Frawley 1991), followed by positioning them within the three cognitive representation levels, i.e. neural, spatial and symbolic (Gärdenfors 2000). As most human knowledge is stored in the symbolic form, while the most powerful machine learning algorithms take as input spatial representations, this explains a plethora of techniques that transform human knowledge into the spatial representation space.

While both approaches, propositionalization and embeddings, aim to transform data into the tabular data format, the works describing these approaches use different terminology and task definitions, claim to have different goals, and are used in very different contexts. We aim to unify the terminological disorder by defining the main categories of data transformation techniques based on the representation they use approaches they employ. Propositionalization approaches produce tabular data mostly from relational databases but also from a mixture of tabular data and background knowledge in the form of logic programs, networked data, and ontologies. Knowledge stored in graphs can be extracted with, e.g., community detection or graph traversal methods. Relations described with similarity matrices are encoded in a numeric form using matrix factorization. Currently, the most promising approach to data transformations is neural networks-based embeddings that can be applied to texts, graphs, and other entities for which we can define a suitable similarity function. We present several of these techniques, focusing on text mining methods, relational learning, and network analysis. We demonstrate how propositionalization techniques can be merged with deep neural network approaches to produce highly efficient joint embedding.

The rest of this monograph is structured as follows. An overview of related machine learning research is presented in Chap. 2. Text embedding approaches are presented in Chap. 3. Structured data representations and related mechanisms for complex feature construction are addressed in the next two chapters. We describe data transformations for mining relational data in Chap. 4, followed by approaches to representation learning from graphs, ontologies, and heterogeneous information networks in Chap. 5. Approaches unifying propositionalization and embeddings into joint data fusion pipelines are presented in Chap. 6. The monograph concludes by discussing the unifying and differentiating aspects of propositionalization and embeddings, as well as their strengths and limitations, in Chap. 7.

References

Rakesh Agrawal and Ramakrishnan Srikant. Fast algorithms for mining association rules in large databases. In *Proceedings of the 20th International Conference on Very Large Data Bases*, pages 487–499, 1994.

Yoshua Bengio, Aaron Courville, and Pascal Vincent. Representation learning: A review and new perspectives. *IEEE Transactions on Pattern analysis and Machine Intelligence*, 35(8):1798–1828, 2013.

David M. Blei, Andrew Y. Ng, and Michael I. Jordan. Latent Dirichlet Allocation. *Journal of Machine Learning Research*, 3:993–1022, 2003.

Hendrik Blockeel, Luc De Raedt, and Jan Ramon. Top-down induction of clustering trees. In *Proceedings of the 15th International Conference on Machine Learning*, pages 55–63. Morgan Kaufmann, 1998.

Leo Breiman. Random forests. *Machine Learning*, 45(1):5–32, 2001.

Leo Breiman, Jerome H. Friedman, R. Olshen, and C. Stone. *Classification and Regression Trees*. Wadsworth & Brooks, 1984.

Peter Clark and Tim Niblett. The CN2 induction algorithm. *Machine Learning*, 3(4):261–283, 1989.

Corinna Cortes and Vladimir Vapnik. Support-vector networks. *Machine Learning*, 20(3):273–297, 1995.

Chad M. Cumby and Dan Roth. On kernel methods for relational learning. In *Proceedings of the 20th International Conference on Machine Learning*, pages 107–114, 2003.

Luc De Raedt. *Logical and Relational Learning*. Springer, 2008.

Usama M. Fayyad, Gregory Piatetsky-Shapiro, Padhraic Smyth, and Ramasamy Uthurusamy, editors. *Advances in Knowledge Discovery and Data Mining*. AAAI Press, Menlo Park, 1995.

Usama M. Fayyad, Gregory Piatetsky-Shapiro, and Padhraic Smyth. From data mining to knowledge discovery in databases. *AI Magazine*, 17(3):37–54, 1996.

Yoav Freund and Robert E. Schapire. A decision-theoretic generalization of on-line learning and an application to boosting. *Journal of Computer and System Sciences*, 55(1):119–139, 1997.

Jerome H. Friedman and Nicholas I. Fisher. Bump hunting in high-dimensional data. *Statistics and Computing*, 9(2):123–143, 1999.

Peter Gärdenfors. *Conceptual Spaces: The Geometry of Thought*. The MIT Press, Cambridge, MA, 2000.

Thomas Gärtner, John W Lloyd, and Peter A Flach. Kernels and distances for structured data. *Machine Learning*, 57(3):205–232, 2004.

Ian Goodfellow, Yoshua Bengio, and Aaron Courville. *Deep Learning*. The MIT Press, 2016.

Jiawei Han and Micheline Kamber. *Data Mining: Concepts and Techniques*. Morgan Kaufmann Publishers, 2001.

David Haussler. Convolution kernels on discrete structures. Technical report, Department of Computer Science, University of California, 1999.

Nada Lavrač and Sašo Džeroski. *Inductive Logic Programming: Techniques and Applications*. Ellis Horwood, 1994.

Jeff Leek. The key word in data science is not data, it is science. *Simply Statistics*, 2013.

David D. Lewis. An evaluation of phrasal and clustered representations on a text categorization task. In *Proceedings of the 15th Annual International ACM SIGIR Conference on Research and Devlopment in Information Retrieval*, pages 37–50, 1992.

Scott M Lundberg and Su-In Lee. A unified approach to interpreting model predictions. In *Advances in Neural Information Processing Systems*, pages 4765–4774, 2017.

Ryszard S. Michalski, Igor Mozetič, Jiarong Hong, and Nada Lavrač. The multi-purpose incremental learning system AQ15 and its testing application on three medical domains. In *Proceedings of the 5th National Conference on Artificial Intelligence*, pages 1041–1045, 1986.

Tomas Mikolov, Ilya Sutskever, Kai Chen, Greg S. Corrado, and Jeff Dean. Distributed representations of words and phrases and their compositionality. In *Advances in neural information processing systems*, pages 3111–3119, 2013.

Tom M. Mitchell. *Machine Learning*. McGraw Hill, 1997.

Stephen H. Muggleton, editor. *Inductive Logic Programming*. Academic Press, London, 1992.

Gregory Piatetsky-Shapiro and William J. Frawley, editors. *Knowledge Discovery in Databases*. The MIT Press, 1991.

Michel Plantié and Michel Crampes. Survey on social community detection. In N. Ramzan et al., editor, *Social Media Retrieval*, pages 65–85. Springer, 2013.

J. Ross Quinlan. Induction of decision trees. *Machine Learning*, 1(1):81–106, 1986.

Marko Robnik-Šikonja and Igor Kononenko. Explaining classifications for individual instances. *IEEE Transactions on Knowledge and Data Engineering*, 20(5):589–600, 2008.

David E. Rumelhart and James L. McClelland, editors. *Parallel Distributed Processing: Explorations in the Microstructure of Cognition*, volume 1: Foundations. The MIT Press, Cambridge, MA, 1986.

David E. Rumelhart, Geoffrey E. Hinton, and Ronald J. Williams. Learning representations by back-propagating errors. *Nature*, 323(6088):533, 1986.

Bernhard Schölkopf and Alexander J. Smola. *Learning with Kernels: Support Vector Machines, Regularization, Optimization, and Beyond*. The MIT Press, 2001.

Vladimir N. Vapnik. *The Nature of Statistical Learning Theory*. Springer, 1995.

Erik Štrumbelj and Igor Kononenko. Explaining prediction models and individual predictions with feature contributions. *Knowledge and Information Systems*, 41(3):647–665, 2014.

David H. Wolpert. Stacked generalization. *Neural Networks*, 5(2):241–260, 1992.

Chapter 2
Machine Learning Background

This chapter provides an introduction to standard machine learning approaches that learn from tabular data representations, followed by an outline of approaches using various other data types addressed in this monograph: texts, relational databases, and networks (graphs, knowledge graphs, and ontologies). We first briefly sketch the historical outline of the research area, establish the basic terminology, and categorize learning tasks in Sect. 2.1. Section 2.2 provides a short introduction to text mining. Section 2.3 introduces relational learning techniques, followed by a brief introduction to network analysis, including semantic data mining, in Sect. 2.4,. The means for evaluating the performance of machine learning algorithms, when used for prediction and rule quality estimation, are outlined in Sect. 2.5. We outline selected data mining techniques and platforms in Sect. 2.6. finally, Sect. 2.7 presents the implemented software that allows for running selected methods on illustrative examples.

2.1 Machine Learning

Machine learning (Mitchell 1997; Hastie et al. 2001; Murphy 2012) is one of the most flourishing areas of computer science. Traditionally, machine learning was concerned with discovering models, patterns, and other regularities in data stored in single data tables. Machine learning research gained additional momentum with the advent of big data and data science, resulting in the need to analyze huge amounts of data gathered in structured relational databases as well as unstructured forms, including texts and images.

In the simplest case, when machine learning algorithms operate on a single data table, the rows in the data table correspond to objects (referred to as *training instances* or *training examples*), and the columns correspond to properties (referred to as *attributes*). The instances are described by *attribute values*, appearing in the

© Springer Nature Switzerland AG 2021
N. Lavrač et al., *Representation Learning*,
https://doi.org/10.1007/978-3-030-68817-2_2

corresponding fields of the data table. In labeled data, i.e. when data instances are labeled by different categories, there is a distinguished attribute named the *class attribute*, whose values correspond to these categories or concepts to which instances belong.

Based on the task addressed, machine learning approaches can be roughly categorized into two main families: supervised and unsupervised learning approaches.

Supervised learning.   This setting, illustrated in Fig. 2.1, assumes that training examples are labeled either by a discrete class label or by a numeric prediction value. A typical learning task is to learn a classification or regression *model* that explains the entire dataset and can be used for classification or prediction of new unlabeled instances. Examples of classification tasks are classification of countries into climate categories, classification of cars into marketing classes based on their properties, or prediction of a diagnosis based on a patient's medical condition. Examples of regression tasks are predicting stock market values, the amount of rainfall in a given period or the number of customers expected in a given day.

Unsupervised learning.   This setting concerns the analysis of unlabeled examples, as illustrated in Fig. 2.2. A typical learning task addressed is to *cluster* the data based on the similarity of instances or to discover descriptive *patterns* that hold for some part of the dataset, describing this part of the data.

Several machine learning tasks reach beyond the definitions of supervised and unsupervised learning given above. For example, *semi-supervised learning* refers

Fig. 2.1 Supervised learning

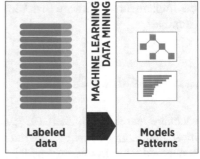

Fig. 2.2 Unsupervised learning

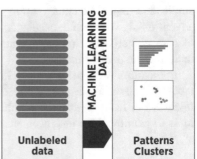

to a learning setting where some training examples are labeled by class labels, and some examples have missing values of the class attribute. Association rule learning algorithms (Agrawal and Srikant 1994; Piatetsky-Shapiro 1991) aim to find interesting relationships between variables, using different measures of interestingness. Subgroup discovery algorithms search for interesting patterns as sets of rules that best describe the target class (Klösgen 1996; Wrobel 1997). *Generative learning* tries to learn the underlying distribution of data and generate new instances from it (Goodfellow et al. 2014; Robnik-Šikonja 2016). *Reinforcement learning* aims to learn actions in the environment that maximize a cumulative reward. The tasks introduced in this paragraph are not discussed in this monograph.

On the other hand, further analysis of supervised learning tasks demonstrates that supervised learning addresses several different learning settings, some of which are relevant in the context of this monograph.

Binary class learning. The simplest setting is referred to as *concept learning* or *binary classification*, where the class attribute is binary, and each training example is labeled with one of the two discrete class values.

Multiclass learning. *Multiclass learning* refers to the setting where the class attribute has more than two discrete class values, while every training example belongs to a single class, and the goal of the learned classifier is to predict the correct class.

Multi-label learning. A broader classification setting is *multi-label learning*, where each training example is labeled with several class labels (or zero class labels). This relaxes the mutual-exclusiveness of the class values assumed in binary and multiclass learning. The *hierarchical multi-label learning* setting is a special case of a multi-label learning, where the set of class labels is organized in a hierarchical fashion using a partial ordering relationship $\leq_R$: if two class labels l_1 and l_2 are in the $l_1 \leq_R l_2$ relationship (e.g., denoting that l_2 is more general than l_1), then a classifier that predicts l_1 should label the instance with class label l_2 as well.

Multi-target learning. In contrast with standard supervised learning, which tries to learn a single target variable at a time, *multi-target learning* (also *multi-target prediction*) tries to learn several related target attributes simultaneously with a single model, hoping to achieve a transfer of knowledge between the tasks. This approach exploits the commonalities and differences across tasks to reduce overfitting and improve generalization. An example of multi-target learning is the prediction of several related medical conditions for a single patient. If the related tasks are described in separate datasets, the learning task is commonly referred to as *multi-task learning*. Further variants of multi-target learning, along with the differences and similarities between them, are discussed by Waegeman et al. (2019).

2.1.1   Attributes and Features

To apply a standard machine learning algorithm, the data should be represented in a single tabular data format, where each row represents a single data instance, and each column represents an attribute or a feature. For the sake of clarity, let us distinguish between *attributes* and *features*.

The attributes that describe the input data instances can be either numeric variables (with values like 7 or 1.5) or nominal/discrete variables (with values like *red* or *female*). In contrast to attributes, features describe the presence or absence of certain properties of an instance. They are always Boolean with values 1 or 0 (*true* or *false*), and usually do not have missing or unknown values. For example, for attribute *gender* there are two features: f_1: *gender=female* and f_2: *gender=male*, and only one of these features is assumed to be *true* for an individual data instance. Features are different even from binary-valued attributes: e.g., for a binary attribute a_i with values *true* and *false*, there are two corresponding features: $a_i = true$ and $a_i = false$. Features can test a value of a single attribute, such as $a_j > 3$, or can represent complex logical and numeric relations, integrating properties of multiple attributes, such as $a_k < 2 \cdot (a_j - a_i)$.

While most feature types only involve a single attribute and can thus be directly mapped to propositional features, relational features differ because they relate the values of two (or more) different attributes to each other. In the simplest case, one could, for example, test for the equality or inequality of the values of two attributes of the same type, such as *Length > Height*. Complex relational features not only make use of the available attributes but may also introduce new variables. For example, if relations are used to encode a graph, a relation like *link(CurrentNode,NextNode)* can subsequently test the neighbor nodes' properties, i.e. nodes connected to the encoded node in the graph.

2.1.2   Machine Learning Approaches

Early machine learning approaches were able to handle only the data represented in a single data table format. Early algorithms include perceptrons (later referred to as neural networks, Rumelhart and McClelland 1986), decision tree learners like ID3 (Quinlan 1979, 1986) and CART (Breiman et al. 1984), and rule learners like AQ (Michalski 1969; Michalski et al. 1986) and INDUCE (Michalski 1980). These algorithms typically learn classifiers from a relatively small set of training examples (up to a thousand) described by a small set of attributes (up to a hundred). An overview of early work in machine learning can be found in Michalski et al. (1983) and Langley (1996). Machine learning approaches can be roughly categorized into two different groups: symbolic and statistical approaches.

Symbolic approaches.   Inductive learning of symbolic descriptions, such as rules (Michalski et al. 1986; Clark and Niblett 1989; Cohen 1995), decision trees (Quinlan 1986) or logical representations (Muggleton 1992; Lavrač and Džeroski 1994; De Raedt 2008).

Statistical approaches.   Statistical or pattern-recognition methods, including k-nearest neighbor or instance-based learning (Dasarathy 1991; Aha et al. 1991), Bayesian classifiers (Pearl 1988), neural network learning (Rumelhart and McClelland 1986), and support vector machines (Vapnik 1995; Schölkopf and Smola 2001).

Since machine learning is a very active research area, many works have progressed the above initial efforts. The textbooks that focus more on symbolic machine learning and data mining research include (Mitchell 1997; Langley 1996; Kononenko and Kukar 2007; Witten et al. 2011). On the other hand, the textbooks dominantly covering statistical learning include (Bishop 2006; Duda et al. 2000; Hastie et al. 2001; Murphy 2012; Goodfellow et al. 2016).

2.1.3   Decision and Regression Tree Learning

A *decision tree* is a classification model whose structure consists of a number of *nodes* and *arcs*. In the simplest case, a node is labeled by an attribute, and its outgoing arcs correspond to the values of this attribute. The top-most node is the *root* of the tree, and the bottom nodes are the *leaves*. Each leaf is labeled by a prediction (i.e. a value of the class attribute).

When used for the classification of an unlabeled instance, the decision tree is traversed in a top-down manner, following the arcs corresponding to the attribute values of the instance for a given attribute in each node. The traversal of the tree leads to a leaf node, and the instance is assigned the class label of the leaf.

A decision tree is constructed in a top-down manner, starting with the root node and then refining it with additional nodes. The crucial step in decision tree induction is the choice of attributes in the tree nodes. A common attribute selection criterion is to use a function that measures the *purity* of class distribution in a node, i.e. the degree to which the node contains instances of a single class. The purity measure is computed for a node and all successor nodes that result from using the attribute in the node to split the training data. For example, the ID3 and C4.5 decision tree algorithms use the entropy as the purity measure (Quinlan 1986). In this algorithm, the difference between the purity of the node and the sum of the purity scores of the successor nodes is weighted by the relative sizes of these nodes and used to estimate the utility of the attribute used in the node. The attribute with the largest utility is selected for tree expansion. The expansion typically stops when the number of instances in a node is low or when all instances are labeled with the same class value. Either way, the resulting leaf bears the label of the most frequent class.

As a result of recursive partitioning of the data at each step in the top-down tree construction process, the number of examples that end up in each successive node steadily decreases. Consequently, the statistical reliability of the chosen attributes decreases with the increasing depth of the tree. As a result, overly complex models may be generated, explaining the training data but not generalizing well to unseen data. This is known as *overfitting*. This is the reason why decision tree learners often employ a post-processing phase in which the generated tree is simplified by *pruning* the branches and nodes near the tree leaves, where they are no longer statistically reliable.

Note that the paths in decision trees can be interpreted as classification rules: the individual paths leading to each of the leaves correspond to individual classification rules. Consequently, decision tree learning can be interpreted as a top-down approach to learning classification rules discussed in the next section.

For regression problems, a very similar procedure to the one described above exists and produces regression trees. Here, the splitting criterion in the nodes is not based on the purity of the class value distribution but the variance of the prediction values. For example, a split on a good attribute would separate instances with large and small prediction values, thereby reducing their variance. The leaves of the tree are typically labeled with the mean of prediction values of the training instances in each leaf.

Many decision and regression tree induction algorithms exist, the most popular being CART (Breiman et al. 1984), C4.5 (Quinlan 1993), and M5 (Quinlan 1992).

2.1.4   Rule Learning

One of the established machine learning techniques is *classification rule learning* (Fürnkranz et al. 2012). Rule learning was initially focused on learning predictive models consisting of a set of classification rules of the form

$$TargetClass \leftarrow Explanation,$$

where the explanation is a logical conjunction of features (attribute values) characterizing the given class. Research in descriptive rule learning includes *association rule learning* (Agrawal and Srikant 1994), which aims at finding interesting descriptive patterns in the unsupervised as well as in supervised learning settings (Liu and Motoda 1998).

Building on classification and association rule learning, *subgroup discovery* techniques aim at finding interesting patterns as sets of rules that best describe the target class (Klösgen 1996; Wrobel 1997). Similar to symbolic rule learning, subgroup discovery techniques also build classification rules; however, each rule shall describe an interesting subgroup of target class instances. Therefore, the main difference between the approaches is in rule selection and model evaluation criteria, briefly addressed in Sect. 2.5.

2.1.5   Kernel Methods

As emphasized in this monograph, the representation space given by the raw data is often insufficient for successful learning of predictive models. In such cases, the raw data representation has to be explicitly or implicitly transformed into a more informative feature vector representations. To perform this transformation, a class of kernel methods requires only a similarity function (a kernel function) over pairs of data points in the raw representation. Let $\phi(x)$ be a transformation of instance x into the feature space, then the kernel function on instances x_1 and x_2 is defined as follows.

$$k(x_1, x_2) = \phi(x_1)^T \cdot \phi(x_2).$$

Here both $\phi(x_1)^T$ and $\phi(x_2)$ are feature vectors, where $\phi(x_1)^T$ is the transpose of $\phi(x_1)$. The simplest kernel is a linear kernel defined as $\phi(x) = x$, in which case $k(x_1, x_2) = x_1^T \cdot x_2$ is the dot product of the two feature vectors. Another family of popular kernels are Radial Basis Functions (RBF) which use the negative distance between arguments as the magnitude of similarity. RBF kernel is typically defined as $k(x_1, x_2) = f(-||x_1 - x_2||^2)$. The most frequent choice is the Gaussian kernel defined as

$$k(x_1, x_2) = \exp(-\gamma ||x_1 - x_2||^2),$$

where $||x_1 - x_2||$ is the Euclidean distance and γ is a parameter.

Kernel functions allow learning in a high-dimensional, implicit feature space without computing the explicit representation of the data in that space. Instead, kernel functions only compute the inner products between the instances in the feature space. This implicit transformation is often computationally cheaper than the explicit computation of the representation. The idea of this kernel trick is to formulate the learning algorithm in such a way that it uses only the similarities between objects formulated as the inner product (i.e. the dot product or scalar product), an algebraic operation that takes two equal-length sequences of numbers (i.e. coordinate vectors) and returns a single number computed as the sum of the products of the corresponding entries of the two sequences of numbers.

Kernel functions have been introduced for vectors (i.e. tabular data), sequence data, graphs, text, images, and many other entities. The most popular algorithms working with kernels are Support Vector Machines (SVMs) (Vapnik 1995). Several other algorithms exist that can work with kernel representation, such as Gaussian processes, Principal Component Analysis (PCA), ridge regression, etc. In the simplest form, the SVM classifier works for binary class problems and finds the maximum-margin hyperplane that divides the instances of the two classes. The margin is defined as the distance between the hyperplane and the nearest point of either class. By using various kernels, SVMs learn non-linear classifiers as well.

2.1.6   Ensemble Methods

Ensemble methods are a well-established approach to predictive learning. We can separate two kinds of approaches, ensembles of homogeneous and ensembles of heterogeneous predictors. Ensembles of homogeneous predictors typically combine a multitude of tree-based models such as decision or regression trees. Ensembles of heterogeneous predictors combine various successful predictors for a given problem, e.g., neural networks, random forests, and SVMs. Homogeneous ensembles can be further split into averaging methods and boosting approaches. Averaging ensembles, such as bagging (Breiman 1996) and random forests (Breiman 2001), build in parallel a multitude of independent dissimilar predictive models and average their output. Boosting (Freund and Schapire 1996), on the other hand, sequentially adds new ensemble members to reduce the errors of previous members.

Note that in this section we use the terms averaging and voting interchangeably as synonyms, as the latter is a special case of averaging for discrete class attributes. While heterogeneous ensemble methods can use voting to obtain the final output, stacking (Wolpert 1992) is a much more successful strategy. Stacking captures the output of several heterogeneous learners (e.g., the predicted probability distribution of class values). It forms a dataset from them, keeping the original example labels. On this dataset, it trains a meta-model that outputs a final prediction. In this way, the certainty of individual models and their success is captured in the final model.

The success of averaging ensemble methods is usually explained with the margin and correlation of base classifiers (Schapire et al. 1997; Breiman 2001). To have a good ensemble, one needs base classifiers that are diverse (in the sense that they predict differently) yet accurate. The voting mechanism, which operates on top of the base learners, then ensures highly accurate predictions of the ensemble.

The original boosting algorithm AdaBoost by Freund and Shapire (1996) constructs a series of base learners by weighting their training set of examples according to the correctness of the prediction. Correctly predicted examples have their weights decreased, and incorrect predictions result in increased weights of such instances. In this way, the subsequent base learners receive effectively different training sets and gradually focus on the most problematic instances. Usually, tree-based models such as decision stumps are used as the base learners.

A recent, highly successful variant of boosting is extreme gradient boosting XGBoost (Chen and Guestrin 2016). Like other boosting variants, it also iteratively trains an ensemble of decision trees but uses a regularized objective function and subsampling of attributes, which help control overfitting. The stated optimization problem is solved with gradient descent. XGBoost can be used for learning in classification, regression, or ranking, depending on the objective function used. Besides good performance, this approach can be paralleliz'ed and is highly scalable.

Random forests (Breiman 2001) construct a series of decision tree-based learners. Each base learner receives a different training set of n instances, drawn independently with replacement from the learning set of n instances. The bootstrap replication of training instances is not the only source of randomness. In each node of the tree, the splitting attribute is selected from a randomly chosen sample of attributes. Random forests are computationally effective and offer good prediction performance. They are proven not to overfit and are less sensitive to noisy data compared to the original boosting. As the training sets of individual trees are constructed by bootstrap replication, there is, on average, $1/e \approx 1/2.718 \approx 36.8\%$ of instances not taking part in the construction of each tree. These instances, named out-of-bag instances, are the source of data for useful internal estimates of error, attribute evaluation, and outlier detection.

Recently, random forests have been used in a deep learning framework, where the learning algorithm constructs a sequence of random forests where later forests correct the errors of the previous layers. The gcForest algorithm (Zhou and Feng 2017) builds a deep forest with a cascade structure, which enables representation learning by forests. The cascade levels are determined automatically based on the dataset. The gcForest is competitive to deep neural networks on many problems but requires fewer data and fewer hyper-parameters.

2.1.7 Deep Neural Networks

Machine learning has relatively recently achieved breakthroughs in terms of solving previously unsolvable problems in computer vision (e.g., face recognition rivals humans' performance), game playing (e.g., a program has beaten human champions in the game of Go), and natural language understanding (e.g., excellent automatic voice recognition and machine translation results). These successes are mostly due to advances in *Deep Neural Networks* (DNNs) (Goodfellow et al. 2016). Artificial neural networks consist of large collections of connected computational units named artificial neurons, loosely analogous to neurons in the brain. Recently, researchers can effectively add more layers of 'neurons' to the networks. To train deep neural networks with many layers, we require large collections of solved examples and fast parallel computers, typically Graphics Processing Units(GPUs).

To understand the main ideas of DNNs, let us start with a neuron and then show the power of their organization into networks. We will only give a minimally sufficient background in this monograph. We refer the interested reader to other relevant sources, such as the monograph by Goodfellow et al. (2016).

A single artificial neuron, shown in Fig. 2.3, builds upon an analogy with biological neurons in the brain. The analogy is shallow: the artificial neuron gets inputs from its neighbors and sums them together, taking the weights of each input into account. The computed sum is an input to the activation function that outputs the values.

Typically, the activation function is analogous to the excitement level of a biological neuron. If the sum exceeds the threshold, the neuron outputs a high value (e.g., 1); otherwise, the output stays low (e.g., value is 0). A few frequently used activation functions are presented below.

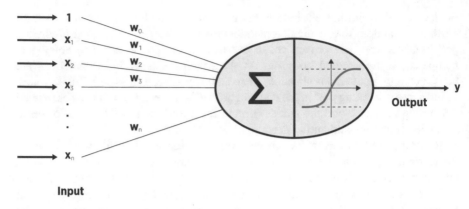

Input

Fig. 2.3 A single artificial neuron

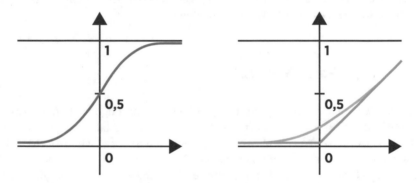

Fig. 2.4 Examples of different activation functions for neural networks: the sigmoid (red) function (illustrated by a graph on the left-hand side), and the ReLu (green) and softplus (grey) functions (illustrated by graphs on the right-hand side)

Sigmoid   is shown in red in Fig. 2.4 and is defined as:

$$f(x) = \frac{1}{1 + e^{-x}}. \tag{2.1}$$

ReLU   (Rectified Linear Unit) is shown in green in Fig. 2.4 and is defined as:

$$f(x) = \max\{0, x\}. \tag{2.2}$$

Softplus   is shown in gray in Fig. 2.4 and is defined as:

$$f(x) = \ln(1 + e^x). \tag{2.3}$$

ELU   (Exponential Linear Unit) (Clevert et al. 2016) is defined as:

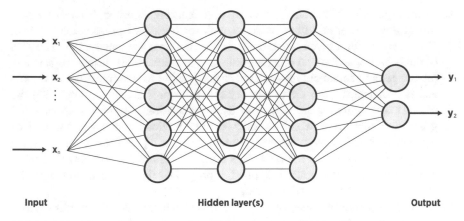

Input   Hidden layer(s)   Output

Fig. 2.5 An illustration of neural network with input layer, three hidden layers, and an output layer

$$ELU(x) = \begin{cases} c \cdot (e^x - 1), & \text{for } x < 0 \\ x & \text{for } x \geq 0 \end{cases} \tag{2.4}$$

where c is the user-specified constant. ELU is similar to ReLU except for negative inputs, where ELU smoothly approaches $-c$, whereas ReLU is zero.

The functioning of single neurons is similar to logistic regression. However, their computational power significantly increases when we connect many neurons in a neural network, shown in Fig. 2.5. In the simplest form of artificial neural networks, referred to as the feedforward networks, neurons are organized into a progressive sequence of layers. The first layer, i.e. the so-called input layer, represents the input to the networks. The last layer, i.e. the output layer, reports the computation results. At least one but possibly many intermediate layers, referred to as the hidden layers, transform the input into the output. For example, a network's input can be pixels forming a black and white image (0 for white and 1 for black color). The outputs may represent probabilities of different digits. Each output is a numeric value between 0 and 1, denoting the probability that the image contains a certain digit, e.g., 9. In feedforward networks, the order of computation is directed and the connections do not form any cycles.

For non-linear activation functions, Hornik et al. (1989) have shown that neural networks can approximate any continuous function (possibly using exponentially many neurons). If a network contains more than one hidden layer, it is referred to as a *deep neural network*. Theoretically, there is no difference in expressive power between shallow and deep neural networks. However, the practice has shown that deep neural networks can be much more successful.

In neural networks, the information is stored in their weights. For neural networks to give any sensible output, they have to be trained, i.e. their weights have to be set in such a way that the inputs are mapped to desired outputs. The most frequently used learning algorithms are backpropagation and recurrent neural networks, both outlined below.

Backpropagation.   The most frequently used learning algorithm is *backpropagation* (Rumelhart et al. 1986), which stands for backward propagation of error. Consider that for a given input, the network produces a wrong output. The idea of the backpropagation learning algorithm is that error—the difference between the correct and actual output—is used as a guideline for modification of weights and is propagated from the output backward towards the input, modifying the network's weights on the way. The backpropagation algorithm computes the gradient of the error, which depends on all previous neurons (composed of weights and activation functions). Computing partial derivatives of the error function with respect to the weights, we get the information on how to modify each weight. This procedure is applied to progressively more distant weights by applying the chain rule of derivation.

Recurrent neural networks (RNNs).   As much as feedforward networks are useful, they have difficulties in processing sequences (for example, words in a sentence forming a sequence). The sequences can be of different lengths, and typically there is some dependency between different positions in a sequence (e.g., a verb in a sentence may determine the subsequent choice of nouns). For such cases, we can use RNNs which contain loops. By introducing backward connections, the information from the previous processing steps persists in the network, effectively allowing the network to memorize previous processing, which is well suited for sequences. Figure 2.6 shows an example of a recurrent neural network. To better understand the functioning of recurrent neural networks, we can imagine them in an unrolled form, i.e. as a sequence of normal layers without cycles. RNNs are very effective in processing speech, text, signals, and other sequential data. RNNs also introduce many new challenges; for example, the convergence of learning to stable weights is much slower.

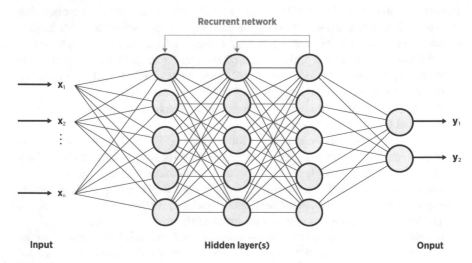

Fig. 2.6 Recurrent neural network

To conclude our short introduction to deep neural networks, we present frequently used types of deep neural networks, i.e. Long Short Term Memory networks (LSTMs), Convolutional Neural Networks (CNNs), and transformers. A common neural network architecture puts a few LSTM or CNN layers at the beginning of a network, followed by one or more fully connected layers. Transformers are a recent architecture, which enables the processing of sequences without recurrent connections by using the attention mechanism that effectively enables longer memory and allows deeper networks.

Long Short Term Memory (LSTM) networks.   The most popular type of RNNs are LSTM networks (Hochreiter and Schmidhuber 1997). LSTM networks allow explicit control over which information is preserved and which is forgotten. The network learns the behavior of its own weights (named cell weights) and the weights of three gates, i.e. the input, output, and forget gate. These gates control the flow of information inside a single LSTM neuron. The input gate controls the flow of new values into the cell; the forget gate controls the remaining values in the cell. The output gate controls which cell values participate in the computation of the output of the LSTM neuron. A variant of the LSTM cell, named *Gated Recurrent Units* (GRUs), do not have an output gate which would explicitly control how its output is constructed.

Convolutional Neural Networks (CNNs).   Like other kinds of artificial neural networks, a convolutional neural network (LeCun et al. 1998) has an input layer, an output layer, and various hidden layers. In CNNs, some of the hidden layers are convolutional, using a mathematical convolution operation to pass on results to successive layers. This simulates some of the actions in the human visual cortex and is very useful for applications in image and video recognition, but also for recommender systems, text processing, and recently also for graphs.

A CNN layer is composed of several learned filters, i.e. detectors for specific data patterns, for example, in an image or text. The behavior of these filters is equivalent to convolutional operators. For pictorial data, CNNs implement sliding of a two-dimensional filter over the whole image and matching the contents of a filter to shapes in the image. This process is illustrated in Fig. 2.7.

After the convolution is computed, the next layer in the network is typically a pooling layer, which aggregates the results of individual matches in a given locality, e.g., 4 × 4. The max-pooling returns the best match (the maximal value of applying the convolutional filter to any image position in the locality). The average pooling returns the average matching of a filter to a local sub-image. This reduces the dimensionality of the input image and helps achieve some position independence (e.g., it does not matter if the pictorial feature such as a bird's beak is exactly at a certain position). For texts, one-dimensional convolutional filters are typically used. They match sequences of characters (e.g., suffixes or stems of important words), as illustrated in Fig. 2.8.

The advantage of using CNNs is that filters are learned from the data, so there is no need to manually engineer them based on the knowledge of which patterns might be important.

Fig. 2.7 An illustration of a
filter applied to an image, as
implemented with CNNs

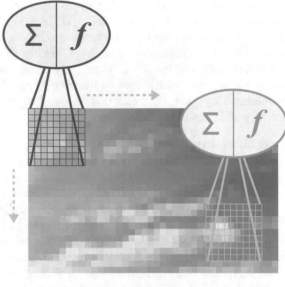

Fig. 2.8 Convolutional filter
for text

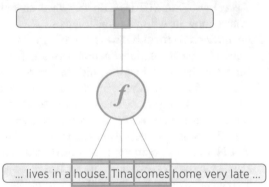

Transformer neural networks.   Transformer networks (Vaswani et al. 2017) are
designed for sequential data, in particular texts. They extensively use the
attention mechanism, originally introduced in machine translation, to resolve
problems with vanishing gradients in backpropagation learning due to long
dependencies. The attention mechanism is essentially an additional memory
used to learn which parts of the previous input are important for the following
ones. Unlike RNNs, transformers do not process the data sequentially. Instead,
they take longer fixed inputs and explicitly encode the sequence order. This
feedforward architecture of transformers allows for better parallelization than
RNNs, shorter training times, and larger datasets. This has led to the development
of large pretrained language models such as BERT (discussed in Sect. 3.3.3),
which have been trained with huge general language datasets, and can be fine-
tuned to specific language tasks.

The transformer is an encoder-decoder architecture. The encoder part consists of a set of encoding layers that processes the input iteratively, one layer after another, producing the internal numeric representation of the (symbolic) input. The decoder consists of a set of decoding layers that decodes the output of the encoder to the output space, e.g., classification, text, or time series. Encoder and decoder layers use the attention mechanism, which for each input, weighs the relevance of every other input, combines them with the input, and produce the output.

Due to their successes, deep neural networks became prevalent machine learning methods for learning from text, images, speech, and graphs. However, deep neural networks are not without problems, e.g., they require large datasets. With a huge number of parameters, they are prone to overfitting. The learning may be slow and requires large amounts of computational and memory resources. Embeddings of discrete data into a numeric space with reduced dimensionality address a part of these problems. First, by reducing the dimensionality of the input data, they reduce the number of neurons and make overfitting less likely. Second, by condensing the sparse information, they reduce the required depth of deep neural networks, thereby speeding up learning and reducing memory requirements.

2.2 Text Mining

Text mining (Feldman and Sanger 2006) is a research area that deals with the construction of models and patterns from text resources, aiming at solving tasks such as text categorization and clustering, taxonomy construction, and sentiment analysis. This research area, also known as text data mining or text analytics, is usually considered a subfield of data mining research (Han and Kamber 2001). It can also be viewed more generally as a multidisciplinary field drawing its techniques from data mining, machine learning, natural language processing (NLP), information retrieval (IR), information extraction (IE), and knowledge management. A typical text mining process is illustrated in Fig. 2.9.

In the data preparation phase, which is in text mining referred to as the *preprocessing step*, the goal is to take an unstructured set of documents and convert it into a table of instances. Several methods embedding text into numeric vectors are presented in Chap. 3.

Historically, in the vector space model (Salton et al. 1975), each document feature represents one attribute of a document. In the preprocessing step, each document is represented as a set of document features and converted into a sparse vector representation. Each row in the constructed data table represents one document.

If a document is unstructured, its features might be derived through preprocessing methods alone. Still, features may also be assigned to it, such as author, keywords,

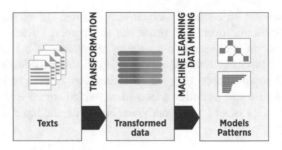

Fig. 2.9 Text mining process

or publication date. Document features may be *words*, *terms*, or *concepts* that occur in a given document.

Words   are the simplest features, extracted and tokenized from the text, bearing the least semantics.

Terms   are sets of one or more words that occur together after being normalized, filtered and grouped through various term extraction methods.

Concepts   are the most semantic-loaded kind of features: they include terms, but also related notions not necessarily mentioned in a document (for example, a document describing an object, such as a computer, might not make use of the word 'computer'; however, the concept 'computer' may be assigned to such document as a feature).

Document features based on words are usually *sparse* (Feldman and Sanger 2006). This is because each document contains only a few features of the entire document corpus.

Traditionally, high-quality features, i.e. features that are informative for a given task, were hard to acquire. Classical approaches countered this by incorporating background knowledge about the documents and their domain. This was achieved by using a controlled vocabulary composed of all terms deemed relevant for a given domain. The modern approaches, presented in Chap. 3, drop the need for manual construction of features and build dense word and document representations.

Standard (pre-neural) approaches to text mining needed long (and often fragile) pipelines of several different processing steps listed below.

Tokenization   splits text into small meaningful units such as words.

Stop-word removal   removes stop-words that are deemed irrelevant for the given task, e.g., prepositions, which are part of every document, are irrelevant for finding similar documents.

Lemmatization   is the process of converting one or several inflected forms of a word into a single, 'morphologically neutral' unit, referred to as a *lemma*). This process allows that different forms of a word are considered semantically identical, at least for a given task. For example, the words 'discover', 'discovering' and 'discovered' all correspond to the lemma 'discover'. An alternative to lemmatization is *stemming*, where inflected forms of a word are truncated to the word stem.

Part-of-Speech (POS) tagging   is the process of annotating words according to their POS category (e.g., noun, verb), based on the context in which they occur.

Dependency parsing   extracts a dependency parse of a sentence that represents its grammatical structure and defines the relationships between 'head' words and words, which modify those heads.

Syntactic parsing   analyzes sentences syntactically according to a specific grammar.

For some text mining algorithms, text data needs to be structured as a table of instances and attributes. Some key text mining concepts include the notions of *corpus*, *document*, and *features*. A corpus, also named *document collection*, is a set of text documents from which patterns can be uncovered. Corpora can vary greatly in size, from a few hundred to millions. A larger corpus is more likely to yield reliable patterns.

We discuss different sparse and dense text representations in Chap. 3, but their common output is the representation of a given text unit (a word, sentence, paragraph, or document) with a numeric vector. Using vector representations of text units, we can compute the similarity of two units, x and y, using the cosine similarity measure, defined in Eq. (3.2).

2.3   Relational Learning

While standard machine learning and data mining approaches learn models or patterns from a single data table, some data analysis scenarios are more complex and aim at finding models or patterns from data stored in multiple tables, e.g., a given relational database. Such relational learning scenarios (Lavrač and Džeroski 1994; Džeroski and Lavrač 2001) require the use of specific learners that can handle multi-tabular data.

Propositional data representation, where each instance is represented with a fixed-length tuple, is natural for many machine learning problems. However, where there is no natural propositional encoding for the problem at hand, some problems require a *relational representation*. For example, it is hard to represent a citation in propositional (tabular) form without losing information since each author can have any number of co-authors and papers. This problem is naturally represented using multiple binary relations, e.g., by using two distinct *author* and *paper* relations.

Consider Table 2.1 that shows a simple relational database schema for storing authorship information about researchers, their fields of research, and papers. Suppose we are interested in finding descriptions of researchers from the field of data mining. An example result of a relational learning algorithm could be as follows:

Table 2.1 An example
relational database schema.
Underlined attributes denote
the private and foreign keys
connecting the tables

researcher	author	paper
researcherID	researcherID	paperID
researcherName	paperID	paperTitle
researchField		conference

researcher(Researcher, 'Data Mining') ← author(Researcher,Paper) ∧
paper(Paper, 'ECML/PKDD').

This logical rule is interpreted as follows: For every researcher, if there exists
a paper authored by the researcher that was published at the ECML-PKDD
conference, then the researcher is in the field of data mining. In the above rule, ← is
a logical implication, and ∧ is the conjunction operator. Relation names *researcher*,
author, *paper* are written by lowercase letters, variable names *Researcher*, *Paper*
start with a capital letter, while constants start with a lowercase letter or by any
letter type in case of using quotes, e.g., *'Data Mining'*, *'ECML/PKDD'*. The logical
implication ← connects the rule head (at the left-hand side of the rule, consisting
of a single literal) and the rule body (at the right-hand side of the rule, consisting of
a conjunction of literals). Here, a literal is an atomic formula $p(t_1, \cdots , t_n)$, where
p is a predicate symbol and its arguments t_i are terms, constituted from constants,
variables, and/or function symbols.

The body of this rule yields researchers that have published papers at the
ECML/PKDD conference, which the algorithm determined as a good pattern for
describing data mining researchers. The rule references the *author*, *paper* and the
main *researcher* table via foreign keys as illustrated in Table 2.1. In this way, the
algorithm exploits the *structural* information available for the learning examples
(i.e. researchers), which illustrates the main powerful feature of relational learning.

The research community has developed numerous approaches to Relational
Learning (RL) (Quinlan 1990), Relational Data Mining (RDM) (Džeroski and
Lavrač 2001), where the discovered patterns/models are expressed in relational
formalisms, and Inductive Logic Programming (ILP) (Muggleton 1992; Lavrač and
Džeroski 1994), where the discovered patterns/models are expressed in first-order
logic.

Since most real-world datasets are stored in some Relational Database Management System (RDBMS), using the relational database format is practically
important, which led to many relational learning algorithms, including the early
relational learning algorithm FOIL (Quinlan 1990). On the other hand, the ILP
community developed several algorithms for learning clausal logic, including
Progol (Muggleton 1995) or Aleph (Srinivasan 2007). The most widely used ILP
system is Aleph (A Learning Engine for Proposing Hypotheses). Some of Aleph's
popularity is due to the fact that the system was conceived as a workbench for
implementing and testing of concepts and procedures from a variety of different
ILP systems and papers, and is described in an extensive user manual of Aleph
(Srinivasan 2007).

Table 2.2 A sample propositional representation of the *researcher* table

researcher					
researcherID	q_1	q_2	...	q_m	researchField
R_1	1	1	...	1	F_1
R_2	0	1	...	0	F_1
R_3	1	0	...	0	F_2
...	...	...	...	...	...
R_n	0	1	0	0	F_1

While the family of relational learning and ILP learning algorithms, including the most popular algorithms, FOIL or Aleph, build a model or induce a set of patterns directly, learning can also be performed indirectly, using a two-step approach named propositionalization.

Propositionalization. In propositionalization, learning is performed in two steps. First, by constructing complex relational features, transforming the relational representation into a propositional single-table format. Second, by applying a propositional learner on the transformed single-table representation.

To illustrate propositionalization, consider a simplified relational problem, where the input data is a database of authors and their papers, presented in a relational database (see the database schema in Table 2.1), where the task is to assign a research field to unseen authors. In essence, complete propositional representation of the problem is obtained by constructing a set of complex relational features and applying them as queries q_i to the target data table, returning value *true* or *false* for a given author. This example is illustrated in Table 2.2. Each query describes a property of a researcher. The property can involve a rather complex query, involving multiple relations, as long as that query returns either *true* or *false*, or the result of some other aggregation function. For example, a query could be 'does author X have a paper published at the ECML/PKDD conference?' or 'how many papers has author X published at the ECML/PKDD conference?'.

Note that propositional representations (a single table format) impose the constraint that each training example is represented as a single fixed-length tuple; the transformation into a propositional representation can be done with many machine learning or data mining tasks in mind: classification, association discovery, clustering, etc. However, propositionalization, which changes the representation for relational learning into a single-table format, cannot always be done without loss of information.

On the one hand, propositionalization is a powerful method when the problem at hand is *individual-centered* (Flach and Lachiche 1999), involving one-to-one or one-to-many relationships. Such problems have a clear notion of an individual. The learning occurs only at the level of (sets of) individual instances rather than the (network of) relationships between the instances, which occur in the case of many-to-many relationships. For example, in the above example problem of classifying authors into research fields given a citation network, the author is an individual, and learning occurs at the author level, i.e. assigning class labels to authors.

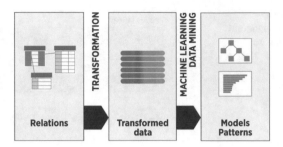

Fig. 2.10 Relational learning, including a propositionalization step involving automated relational data transformation

On the other hand, for some relational problems, there may not exist an elegant propositional encoding. For example, in general, a citation network cannot be represented in a propositional format without loss of information since each author can have any number of co-authors and papers. In this case, the problem is naturally represented using multiple relations, e.g., including the *author* and the *paper* relations.

While relational data transformations can be performed manually by a data analyst, we are only interested in automated propositionalization methods. Automated relational data transformation using propositionalization, which is illustrated in Fig. 2.10, is discussed in detail in Chap. 4.

2.4   Network Analysis

Compared to a standard tabular data representation, more complex data representations are needed when instances under examination are interconnected to a varying (non-fixed) number of other instances. Representing each connection from an instance as a separate column would result in a different number of columns for each row. Alternatively, if we encoded connections of an instance with a single column, columns would have to contain composite data structures of varying length, such as lists.

An alternative way to describe a dataset containing inter-connected instances is to represent it as a *network* (Burt and Minor 1983), a data structure containing nodes and connections/relations between the nodes. The field of *network analysis* has its roots in two research fields: mathematical graph theory and social sciences. Network analysis started as an independent research discipline in the late seventies (Zachary 1977) and early eighties (Burt and Minor 1983) when sociologists became increasingly aware that the study of social relations—and not only individual attributes—is necessary for in-depth analysis of human societies. Network analysis uses graph theory concepts to investigate networked structures' characteristics in terms of nodes (such as individual actors, people, or any other objects in the network) and edges (links) that connect them (relationships, interactions, or ties between the objects). Since early network analysis research, which focused on *social networks*, where nodes represent people and connections represent their social links, network analysis has grown substantially. The field now covers not only social networks but also networks originating from many other (scientific) disciplines. For example, biological networks include gene regulatory networks, protein interaction networks, drug interaction networks, etc. Of particular interest are the so-called *information networks*, where the directed connections encode the information flow between the network nodes.

In mathematical terms, network structures are represented by *graphs*, where the nodes are referred to as vertices, and their connections/relations are referred to as edges. Formally, a *graph* $G = (V, E)$ is a mathematical object, composed of a set of vertices (nodes) V, and a set of edges (links) E connecting the vertices. Edges between pairs of vertices (u, v) can be either *directed* or *undirected*, where *undirected graphs* consisting of only undirected edges are considered as a subclass of the class of *directed graphs*.

A graph with no loops (edges connecting a node to itself) and no multiple edges (meaning that a pair of nodes is connected by at most one edge) is a *simple* graph. *Homogeneous graphs* refer to graphs where all the nodes and edges are of the same type. In contrast, *heterogeneous graphs* refer to graphs composed of various types of nodes and edges. Similarly, compared to simpler *homogeneous networks* including *homogeneous information networks*, more complex types are *heterogeneous information networks* that consist of various types of nodes and links, which became a popular field of study due to Sun and Han (2012).

Graph mining and network analysis cover a wide variety of tasks. In this section, we first list some analytic tasks, which can be applied to homogeneous graphs and networks, followed by selected tasks on heterogeneous networks. We then describe semantic data mining, using *ontologies* as the background knowledge, and briefly introduce ontology propositionalization and network embeddings, including *knowledge graph* embeddings.

2.4.1 Selected Homogeneous Network Analysis Tasks

Network node classification.   Classification of network data is a natural generalization of classification tasks encountered in a typical machine learning setting. The problem formulation is simple: given a network and class labels for some of the network entities, predict the class labels for the rest of the entities in the network. Another name for this problem is *label propagation*. Zhou et al. (2004) proposed a commonly used algorithm for this task. For example, the method was used to discover new genes associated with a disease (Vanunu et al. 2010).

Link prediction.   While classification tasks try to discover new knowledge about
network entities, link prediction focuses on unknown connections between the
entities. The assumption is that some network edges are not known. The task
of link prediction is to predict new edges that are missing or likely to appear in
the future. A common approach to link prediction is assigning a score $s(u, v)$
to each pair of vertices u and v, which models a probability of the connected
vertices. Link prediction approaches include calculating the score as a product of
vertex degrees (Barabási et al. 2002) and (Newman 2001a), or using the number
of common neighbors of two vertices $|N_u \cap N_v|$ (Newman 2001b).

Community detection.   While there is no strict definition of the term *network
community*, the idea is well summarized by Yang et al. (2010) as follows: a
community is a group of network nodes, with dense links within the group and
sparse links between the rest of the groups. An extensive overview of community
detection methods is provided by Plantié and Crampes (2013).

Network node ranking.   The objective of ranking in information networks is to
assess the relevance of a given object either globally (concerning the whole
graph) or locally (relative to some object in the graph). A well-known ranking
method is PageRank (Page et al. 1999), which was initially used in the Google
search engine. The idea of PageRank (PR) can be explained in two ways.

- The first is the random walker approach: a random walker starts walking
  from a random vertex v of the network. In each step, it walks to one of the
  neighboring vertices with a probability proportional to the weight of the edge
  traversed. The PageRank of a vertex is then the expected proportion of time
  the walker spends in the vertex or, equivalently, the probability that the walker
  is in the particular vertex after a long time.
- The second view of PageRank is the view of score propagation. The PageRank
  of a vertex is its score, which it passes to the neighboring vertices. A vertex
  v_i with a score $PR(v_i)$ transfers its score to all its neighbors. Each neighbor
  receives a share of the score proportional to the strength of the edge between
  it and v_i. This view considers that for a vertex to be highly ranked, it must be
  pointed to by many highly ranked vertices.

Other methods for ranking include Personalized-PageRank (Page et al. 1999),
frequently abbreviated as P-PR, that calculates the vertex score locally to a given
network vertex, SimRank (Jeh and Widom 2002), diffusion kernels (Kondor and
Lafferty 2002), hubs and authorities (Chakrabarti et al. 1998) and spreading
activation (Crestani 1997).

2.4.2 Selected Heterogeneous Network Analysis Tasks

While most researchers deal with *homogeneous graphs* and *homogeneous information networks*, where all nodes and edges are of the same node/edge type, Sun and
Han (2012) addressed the problem of *heterogeneous information network* analysis,

where nodes and edges belong to different node or edge types. For example, we may have a network containing both the genes and the proteins they encode, which necessitates the use of two node types to represent the data. While most of the approaches developed for analyzing homogeneous information networks can be applied to heterogeneous networks by simply ignoring the heterogeneous network structure, this decreases the amount of information available for data analysis and can decrease its performance (Davis et al. 2011), therefore special approaches that take the heterogeneous network structure into account have been developed. Selected approaches are listed below.

Authority ranking. Sun and Han (2012) introduced authority ranking to rank the vertices of a bipartite network.

Ranking based clustering. Joining two seemingly orthogonal approaches to information network analysis (ranking and clustering) into a joint approach was achieved by Sun and Han (2012) by using two algorithms: RankClus (Sun et al. 2009a) and NetClus (Sun et al. 2009b), both of which cluster entities of a certain type (for example, authors) into clusters and rank the entities within clusters. Algorithm RankClus is tailored for bipartite information networks, while NetClus can be applied to networks with a star network schema.

Classification through label propagation. Hwang and Kuang (2010) expanded the idea of label propagation used for homogeneous networks by Zhou et al. (2004) to find a probability distribution f of a node being labeled with a positive label. Similarly, Ji et al. (2010) proposed the GNETMINE algorithm, which uses the idea of knowledge propagation through a heterogeneous information network to find probability estimates for labels of the unlabeled data.

Ranking based classification. Building on the idea of GNETMINE, Sun and Han (2012) proposed the RankClass classification algorithm that relies on within-class ranking functions to achieve better classification results.

Relational link prediction. Expanding the ideas of link prediction for homogeneous information networks, Davis et al. (2011) proposed a link prediction algorithm for each pair of object types in the network, where the score is higher if the two objects are likely to be linked.

Semantic link association prediction. A more advanced approach to link prediction was proposed by Chen et al. (2012), who developed a statistic model, named Semantic Link Association Prediction (SLAP), to measure associations between elements of a heterogeneous network constructed from 17 publicly available data sources about drug-target interaction, including semantically annotated knowledge sources in the form of ontologies.

2.4.3 Semantic Data Mining

Machine learning approaches can greatly benefit from using domain-specific knowledge, which we refer to as *background knowledge*. The background knowledge is usually structured in the form of taxonomies or some more elaborate knowledge structures curated by human experts. With the advent of the semantic web (Berners-Lee et al. 2001), *ontologies* emerged as the representative data structure for encoding background knowledge. Formally, ontologies are directed acyclic graphs, formed of concepts and their relations, encoded as subject-predicate-object (s, p, o) triplets. These knowledge sources offer valuable additional information, often not accessible to algorithms operating on raw data only.

A data mining scenario in which learning is performed on the data, annotated by concepts from human-curated knowledge sources such as ontologies, is referred to as *semantic data mining*. Given the abundance of taxonomies and ontologies that are readily available, rule learning can be used to provide higher-level descriptors and explanations of groups of instances covered by the discovered rules. For example, in the domain of systems biology, the Gene Ontology (Ashburner et al. 2000), KEGG orthology (Ogata et al. 1999) and Entrez gene-gene interaction data (Maglott et al. 2005) are examples of structured domain knowledge. In rule learning, the terms in the nodes of these ontologies can take the role of additional high-level descriptors (generalizations of data instances) used in the induced rules; the hierarchical relations among the terms can guide the search for conjuncts of the induced rules.

To find patterns in data annotated with ontologies, we rely on Semantic Data Mining (SDM) algorithms (Dou et al. 2015; Lavrač and Vavpetič 2015), where the input is composed of a set of class-labeled instances and the background knowledge encoded in the form of ontologies. The goal is to find descriptions of target class instances as a set of rules of the form *TargetClass ← Explanation*, where the explanation is logical conjunction of terms from the ontology. Semantic data mining has its roots in symbolic rule learning, subgroup discovery, and enrichment analysis research.

Recent achievements in the area of deep learning (LeCun et al. 2015) have also resulted in a series of novel, *semantics-aware* learning approaches to previously well-established problems of, e.g., entity resolution, heterogeneous network embedding, author profiling, recommendation systems, and ontology learning. Ontology propositionalization, illustrated in Fig. 2.11, is addressed in Sect. 5.4.

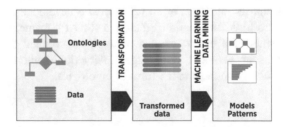

Fig. 2.11 Semantic data mining, including an automated ontology transformation step

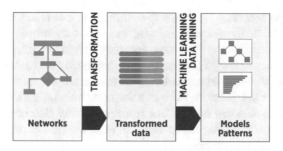

Fig. 2.12 Network analysis, including an automated network transformation step

2.4.4 Network Representation Learning

Network representation learning, addressed in Chap. 5, is the process of transforming a given network, its topology as well as node and edge labels, into a vector representation format (Cui et al. 2018) that can be used in various down-stream learning tasks, as outlined in Fig. 2.12. In the transformed vector representation space, rows correspond to individual network nodes, and the columns correspond to automatically constructed features used for describing the properties of nodes in the transformed representation space. More rarely, rows may correspond to edges and columns to their properties.

A specific form of networks that annotate real-world data are commonly referred to as *knowledge graphs*. Examples of knowledge graphs include DBpedia (Auer et al. 2007), WordNet (Miller 1995), and YAGO (Suchanek et al. 2007). The task of knowledge graph embedding is addressed in Sect. 5.5.

2.5 Evaluation

When evaluating a machine learning prediction model, we typically split the data into two sets, training set and test set, and estimate the models' predictive performance on the test set. For small datasets where a hold-out set would significantly reduce the learning capability due to wasted training data, we use a cross-validation approach. We shortly describe these concepts below, followed by common evaluation metrics: classification accuracy, classification error, precision, recall, and F-measure. We end the section with the rule quality evaluation measures: coverage, support, and lift.

Training set and test set.   It is overly optimistic to estimate the predictive accuracy of a model on the same dataset that was used for learning. To get a realistic performance estimate, the predictive accuracy should be estimated on a separate part of the data, a so-called *test set*, that was removed (held out) from the training phase. While hold-out set evaluation produces more accurate estimates for the predictive accuracy, it also results in a waste of training data because only part of the available data can be used for learning. It may also be unreliable if only a small test set is used due to the variance resulting from a random sampling of training and testing instances.

Cross-validation.   Cross-validation is a common method for easing the problem of insufficiently large training and test sets: while all of the data are used for learning, the accuracy of the resulting model is estimated by performing k hold-out experiments as described above. For this purpose, the data is divided into k parts. In each experiment, k-1 parts are combined into a training set, and the remaining part is used for testing. A machine learning model is then learned on the training set and evaluated on the test set. This is repeated until each part (and thus each example) has been used for testing once. The final accuracy is estimated as an average of the accuracy estimates computed in each such hold-out experiment. The cross-validation procedure can be used for estimating any aspect of the quality of the learned model.

2.5.1   Classifier Evaluation Measures

Let us first define the instances belonging to different classes, and the corresponding confusion matrix. In a binary classification problem, let E denote the set of all training instances, P denote the set of positive instances, and N the set of negative instances, where $P \cup N = E$ and $|P| + |N| = |E|$. Let $TP \subseteq P$ (true positives) be a set of positive instances that are correctly classified by the learned model, $TN \subseteq N$ (true negatives) be a set of correctly classified negative instances, $FP \subseteq N$ (false positives) be a set of negative instances that are incorrectly classified as positives by the learned model, and $FN \subseteq P$ (false negatives) be a set of positive instances incorrectly classified as negative instances. These notions are explained in Table 2.3.

We now present the most frequently used evaluation metrics for classification problems.

Table 2.3 Confusion matrix depicting the notation for sets of actual and predicted positive and negative examples

	Predicted Positive	Predicted Negative	Σ
Actual Positive	# True Positives $|TP|$	# False Negatives $|FN|$	$|P|$
Actual Negative	# False Positives $|FP|$	# True Negatives $|TN|$	$|N|$

Classification accuracy. The Classification Accuracy (CA) is defined as the percentage of the total number of correctly classified instances of all classes relative to the total number of tested instances. In a binary classification problem, the accuracy of a classifier is computed as follows.

$$CA = \frac{|TP| + |TN|}{|E|}. \tag{2.5}$$

Note that the accuracy estimate computes the classifier's accuracy on both positive and negative examples of the target class of interest. Its range is [0, 1], larger values meaning better accuracy.

In the case of multiple classes, we denote the set of all classes with C, and the different classes (i.e. class values) with c_i, $i = 1, \ldots, |C|$. The set of examples belonging to class c_i is denoted by C_i, and individual instances from this class are $C_{i,j}$. The number of examples in class c_i is $|C_i|$, where the number of all examples in the training set is $|E| = \sum_{i=1}^{|C|} |C_i|$. Classification accuracy is defined as the ratio between the total number of correctly classified examples in all classes and the total number of tested examples. For multiclass classification, the accuracy of a model is thus defined as

$$Accuracy = \frac{1}{|E|} \sum_{i=1}^{C} \sum_{j=1}^{|C_i|} I(f(C_{i,j}) = c_i), \tag{2.6}$$

where $f()$ is the prediction of a given model (i.e. a class value), and $I()$ is the indicator function, which returns 1 if its argument is true and 0 otherwise.

Classification error. Instead of accuracy, results are often presented in terms of the *classification error*, computed as follows.

$$Error = 1 - Accuracy. \tag{2.7}$$

Precision, recall, and F-measure. In binary class problems, precision is the ratio between the number of correctly classified positive instances and all predicted as positives, i.e.

$$Precision = \frac{|TP|}{|TP| + |FP|}. \tag{2.8}$$

Recall (also known as sensitivity) in binary problems is the ratio of positive instances to the number of all positive instances.

$$Recall = \frac{|TP|}{|TP| + |FN|}. \tag{2.9}$$

A measure that combines precision and recall in binary problems as a harmonic mean of precision and recall is the F_1 measure or the balanced F-score. It is defined as follows.

$$F_1 = 2 \cdot \frac{Precision \cdot Recall}{Precision + Recall}. \tag{2.10}$$

The range of precision, recall, and F_1 is [0, 1] and larger values mean better performance.

Precision, recall, and F-measure for multiclass problems.   The above definitions of precision, recall, and F-measure can be extended to several problems or more than two classes. In the multiclass extension, we treat each class label as a positive class in turn, and all the other labels as a negative class. This gives us a set of $|C|$ different problems, which we can macro-average and micro-average. Either way, we first compute the confusion matrix for each of the problems separately, obtaining sets TP_i, FP_i, FN_i, and TN_i for problem i.

In macro-averaging we compute precision and recall for each of the problems and then average these scores, while F_1^{macro} is the harmonic mean of macro-averaged precision and recall.

$$Precision^{macro} = \frac{1}{|C|} \sum_{i=1}^{|C|} Precision_i \tag{2.11}$$

$$Recall^{macro} = \frac{1}{|C|} \sum_{i=1}^{|C|} Recall_i \tag{2.12}$$

$$F_1^{macro} = 2 \cdot \frac{Precision^{macro} \cdot Recall^{macro}}{Precision^{macro} + Recall^{macro}}. \tag{2.13}$$

In micro-averaging, we first sum up the individual true positives, false positives, true negatives, and false negatives of different problems and then use them in the definitions of precision and recall, as given in Eqs. 2.8 and 2.9. F_1^{micro} is the harmonic mean of micro-averaged precision and recall.

$$Precision^{micro} = \frac{\sum_{i=1}^{|C|} |TP_i|}{\sum_{i=1}^{|C|} |TP_i| + \sum_{i=1}^{|C|} |FP_i|} \tag{2.14}$$

$$Recall^{micro} = \frac{\sum_{i=1}^{|C|} |TP_i|}{\sum_{i=1}^{|C|} |TP_i| + \sum_{i=1}^{|C|} |FN_i|} \tag{2.15}$$

$$F_1^{micro} = 2 \cdot \frac{Precision^{micro} \cdot Recall^{micro}}{Precision^{micro} + Recall^{micro}} \tag{2.16}$$

Macro-average method is used to assess overall performance across all problems, i.e. classes. Here, all class values are treated equally. On the other hand, micro-averages are useful when the sizes of different classes vary and we want to assess the overall performance, disregarding these differences.

2.5.2 Rule Evaluation Measures

In data mining, including association rule learning and subgroup discovery that are briefly addressed in Sect. 2.6.1, we frequently assess rule quality by computing the coverage and support of each individual rule R, where

$$Coverage(R) = |TP| + |FP| \tag{2.17}$$

is the total number of instances covered by the rule. The support of the rule is calculated as:

$$Support(R) = \frac{|TP| + |FP|}{|E|}. \tag{2.18}$$

Both coverage and support assume binary classification and return values in [0, 1] range, where larger values mean better rules.

A frequently used rule quality measure is Lift(R), which is defined as the ratio between the support of the rule and the expected support of the rule. It is calculated as:

$$Lift(R) = \frac{Support(R)}{\hat{p}}. \tag{2.19}$$

where the expected support is the proportion of positive examples P in dataset E, i.e. $\hat{p} = \frac{|P|}{|E|}$, corresponding to the support of the empty rule that classifies all the examples as positive. The range of *Lift* values is $[0, \infty)$, where the larger values indicate better rules.

To evaluate the whole rule set, we treat the rule set as any other classifier, using evaluation measures such as classification accuracy, introduced in Sect. 2.5.1.

2.6 Data Mining and Selected Data Mining Platforms

This section starts with a brief overview of selected data mining approaches, followed by an outline of selected data mining platforms.

2.6.1   Data Mining

Note that the term *machine learning* is frequently used as a synonym for supervised learning, while the term *data mining* is used as a synonym for pattern mining. For example, rule learning, which was initially focused on building predictive models formed of sets of classification rules, has recently shifted its focus to descriptive pattern mining.

There is a growing popularity of data mining in the so-called pattern mining scenarios. Well-known pattern mining techniques in the literature are based on *association rule learning* (Agrawal and Srikant 1994; Piatetsky-Shapiro 1991). Similarly, mining closed itemsets and maximal frequent itemsets has been popular in the data mining community for unlabeled data. An itemset is closed if none of its immediate supersets has the same support as the itemset (Zaki and Hsiao 2005). An itemset is maximally frequent if none of its immediate supersets is frequent (Burdick et al. 2001). However, closed itemset mining was not used in the analysis of labeled data until Garriga et al. (2008) adapted closed sets mining to the task of discriminating different classes by contrasting covering properties on the positive and negative examples. While the initial studies in association rule mining have focused on finding interesting patterns from large datasets in an unsupervised setting, association rules have also been used in a supervised setting to learn pattern descriptions from class-labeled data (Liu and Motoda 1998).

Building on the research in classification and association rule learning, *subgroup discovery* has emerged as a popular data mining methodology for finding patterns in the class-labeled data. Subgroup discovery aims at finding interesting patterns as sets of individual rules that best describe the target class (Klösgen 1996; Wrobel 1997). Similarly, contrast set mining aims to learn the patterns that differentiate one group of instances from another (Kralj Novak et al. 2009). Additionally, emerging pattern mining algorithms construct itemsets that significantly differ from one class to the other(s) (Dong and Li 1999; Kralj Novak et al. 2009).

2.6.2   Selected Data Mining Platforms

The development of modern programming languages, programming paradigms, and operating systems initiated research on computer platforms for data analysis and the development of modern integrated data analysis software. Such platforms offer a high level of abstraction, enabling the user to focus on the analysis of results rather than on the ways of obtaining them. In the beginning, single algorithms were implemented as complete solutions to specific data mining problems, e.g., the C4.5 algorithm (Quinlan 1993) for the induction of decision trees. Second-generation systems like SPSS Clementine, SGI Mineset, IBM Intelligent Miner, and SAS Enterprise Miner were developed by large vendors, offering solutions to typical data preprocessing, transformation and discovery tasks, and also providing graphical user interfaces (He 2009). Many of the later, still ongoing developments, e.g., Weka (Witten et al. 2011) and scikit-learn (Pedregosa et al. 2011) took advantage of the operating system independent languages such as the Java platform or Python environment to produce complete solutions, which include methods for data preprocessing and visual representation of results. These solutions also provide interfaces to call one another and thereby enable cross-solution development.

Visual programming (Burnett 2001), an approach to programming where a procedure or a program is constructed by arranging program elements graphically instead of writing the program as a text, has become widely recognized as an important element of an intuitive interface to complex data mining procedures. All modern advanced knowledge discovery systems offer some form of workflow construction and execution. This is crucial for conducting complex scientific experiments, which need to be repeatable and easy to verify at an abstract level and through experimental evaluation. The standard data mining platforms like Weka (Witten et al. 2011), RapidMiner (Mierswa et al. 2006), KNIME (Berthold et al. 2009) and Orange (Demšar et al. 2004) provide a large collection of generic algorithm implementations, usually coupled with an easy-to-use graphical user interface. Related to these systems is the cloud-based data mining platform ClowdFlows (Kranjc et al. 2017) that provides access to many modern data mining algorithms. Another venue to simplify access to data science tools and enable cross-solution development are modern web-based Jupyter notebooks (Kluyver et al. 2016) that allow to create and share documents that contain live code, equations, and visualizations. We use this solution to practically illustrate the concepts presented in this monograph.

While the operating system independence and the ability to execute data mining workflows was the most distinguished feature of standard data mining platforms a decade ago, today's data mining software is confronted with the challenge of how to make use of newly developed paradigms for big data and deep learning which require grid and GPU processing.

2.7 Implementation and Reuse

Except for Chap. 1, each chapter of this monographconcludes with a dedicated section, named *Implementation and reuse*, where Jupyter Python notebooks are used to demonstrate a selection of algorithms described in a given chapter, allowing the reader to run a selection of described methods on illustrative examples. The notebooks provide a set of examples that have been carefully chosen to be easy to understand but are not trivial and can be used as recipes or starting points in developing more complex data analysis scenarios.

To use the notebooks, the JupyterLab application needs to be installed and activated. The following steps provide the complete setup of the environment where the notebooks can be inspected, edited, run, and reused.

1. **Install the Python language interpreter**. It is recommended to use the latest version of the 3.8 branch to ensure maximum support and stability (older 3.6 and 3.7 branches were also tested and should work as well). The official repository[1] provides a large selection of installers of different versions for all supported platforms.

2. **Download the monograph's repository of notebooks**. The repository can be downloaded by using the git version control system or by downloading the archive of the selected branch. E.g., the archive of the master branch is available at https://github.com/vpodpecan/representation_learning/archive/master.zip.

3. **Create a virtual environment.** A virtual environment provides an isolated development environment for Python packages. It is recommended to create a separate virtual environment for each notebook to minimize potential conflicts between different versions of the required packages. The following commands create and activate a fresh virtual environment named 'myEnv' in the current folder on a Linux-based operating system:

```
python3 -m venv myEnv
source myEnv/bin/activate
```

On a Windows operating system the corresponding commands are as follows:

```
python3 -m venv myEnv
myEnv\Scripts\activate
```

4. **Install JupyterLab**. JupyterLab is a web-based interactive development environment for Jupyter notebooks, code, and data. You will use it to open, run, and modify notebooks from the repository of this monograph. With the virtual environment activated, the following command installs the latest version of `jupyterlab`:

```
pip install jupyterlab
```

5. **Run JupyterLab**. The JupyterLab web application can be launched with the following command:

```
jupyter lab
```

The command will print a local web address where the JupyterLab web application is running. On most platforms, the page will open automatically in your default web browser.

6. **Open a Jupyter notebook.** JupyterLab allows access to files and subfolders in the folder where it was launched. It is therefore recommended to run Jupyter inside the directory containing the monograph's repository of notebooks.

---

[1]https://www.python.org/downloads/.

References

Rakesh Agrawal and Ramakrishnan Srikant. Fast algorithms for mining association rules in large databases. In *Proceedings of the 20th International Conference on Very Large Data Bases*, pages 487–499, 1994.

David W. Aha, Dennis Kibler, and Marc K. Albert. Instance-based learning algorithms. *Machine Learning*, 6:37–66, 1991.

Michael Ashburner, Catherine A. Ball, Judith A. Blake, David Botstein, Heather Butler, J. Michael Cherry, Allan P. Davis, Kara Dolinski, Selina S. Dwight, Janan T. Eppig, Midori A. Harris, David P. Hill, Laurie Issel-Tarver, Andrew Kasarskis, Suzanna Lewis, John C. Matese, Joel E. Richardson, Martin Ringwald, Gerald M. Rubin, and Gavin Sherlock. Gene Ontology: Tool for the unification of biology. *Nature Genetics*, 25(1):25, 2000.

Sören Auer, Christian Bizer, Georgi Kobilarov, Jens Lehmann, Richard Cyganiak, and Zachary G. Ives. DBbpedia: A nucleus for a web of open data. In *Proceedings of the 6th International Semantic Web Conference*, volume 4825 of *Lecture Notes in Computer Science*, pages 722–735. Springer, 2007.

Albert-Laszlo Barabási, Hawoong Jeong, Zoltán Néda, Erzsébet Ravasz, Andras Schubert, and Tamás Vicsek. Evolution of the social network of scientific collaborations. *Physica A: Statistical Mechanics and its Applications*, 311(3-4):590–614, 2002.

Tim Berners-Lee, James Hendler, and Ora Lassila. The semantic web. *Scientific American*, 284 (5):34–43, May 2001.

Michael R. Berthold, Nicolas Cebron, Fabian Dill, Thomas R. Gabriel, Tobias Kötter, Thorsten Meinl, Peter Ohl, Kilian Thiel, and Bernd Wiswedel. KNIME – The Konstanz information miner. Version 2.0 and beyond. *SIGKDD Explorations*, 11:26–31, 2009.

Christopher M. Bishop. *Pattern Recognition and Machine Learning*. Springer, 2006.

Leo Breiman. Bagging predictors. *Machine Learning Journal*, 26(2):123–140, 1996.

Leo Breiman. Random forests. *Machine Learning*, 45(1):5–32, 2001.

Leo Breiman, Jerome H. Friedman, R. Olshen, and C. Stone. *Classification and Regression Trees*. Wadsworth & Brooks, 1984.

Douglas Burdick, Manuel Calimlim, and Johannes Gehrke. MAFIA: A maximal frequent itemset algorithm for transactional databases. In *Proceedings of the 17th International Conference on Data Engineering, 2001*, pages 443–452, 2001.

Margaret M. Burnett. Visual Programming. In *Wiley Encyclopedia of Electrical and Electronics Engineering*, pages 275–283. John Wiley & Sons, 2001.

Ronald S. Burt and Michael J. Minor. *Applied Network Analysis: A Methodological Introduction*. Sage Publications, 1983.

Soumen Chakrabarti, Byron Dom, Prabhakar Raghavan, Sridhar Rajagopalan, David Gibson, and Jon Kleinberg. Automatic resource compilation by analyzing hyperlink structure and associated text. *Computer Networks*, 30(1–7):65–74, 1998.

Bin Chen, Ying Ding, and David J. Wild. Assessing drug target association using semantic linked data. *PLoS Computational Biology*, 8(7), 2012.

Tianqi Chen and Carlos Guestrin. XGBoost: A scalable tree boosting system. In *Proceedings of the 22nd ACM SIGKDD International Conference on Knowledge Discovery and Data Mining*, page 785–794, 2016.

Peter Clark and Tim Niblett. The CN2 induction algorithm. *Machine Learning*, 3(4):261–283, 1989.

Djork-Arné Clevert, Thomas Unterthiner, and Sepp Hochreiter. Fast and accurate deep network learning by exponential linear units (ELUs). In *Proceedings of the International Conference on Representation Learning, ICLR*, 2016.

William W. Cohen. Fast effective rule induction. In *Proceedings of the 12th International Conference on Machine Learning (ML-95)*, pages 115–123, 1995.

Fabio Crestani. Application of spreading activation techniques in information retrieval. *Artificial Intelligence Review*, 11(6):453–482, 1997.

Peng Cui, Xiao Wang, Jian Pei, and Wenwu Zhu. A survey on network embedding. *IEEE Transactions on Knowledge and Data Engineering*, 31(5):833–852, 2018.

Belur V. Dasarathy, editor. *Nearest Neighbor (NN) Norms: NN Pattern Classification Techniques*. IEEE Computer Society Press, Los Alamitos, CA, 1991.

Darcy Davis, Ryan Lichtenwalter, and Nitesh V. Chawla. Multi-relational link prediction in heterogeneous information networks. In *Proceedings of the 2011 International Conference on Advances in Social Networks Analysis and Mining*, pages 281–288, 2011.

Luc De Raedt. *Logical and Relational Learning*. Springer, 2008.

Janez Demšar, Blaž Zupan, Gregor Leban, and Tomaz Curk. Orange: From experimental machine learning to interactive data mining. In *Proceedings of the 8th European Conference on Principles and Practice of Knowledge Discovery in Databases*, pages 537–539, 2004.

Guozhu Dong and Jinyan Li. Efficient mining of emerging patterns: Discovering trends and differences. In *Proceedings of the 5th ACM SIGKDD International Conference on Knowledge Discovery and Data Mining (KDD-99)*, pages 43–52, 1999.

Dejing Dou, Hao Wang, and Haishan Liu. Semantic data mining: A survey of ontology-based approaches. In *Proceedings of the 2015 IEEE International Conference on Semantic Computing (ICSC)*, pages 244–251, 2015.

Richard O. Duda, Peter E. Hart, and David G. Stork. *Pattern Classification*. John Wiley and Sons, 2nd edition, 2000.

Sašo Džeroski and Nada Lavrač, editors. *Relational Data Mining*. Springer, Berlin, 2001.

Ronen Feldman and James Sanger. *The Text Mining Handbook: Advanced Approaches in Analyzing Unstructured Data*. Cambridge University Press, 2006.

Peter Flach and Nicholas Lachiche. 1BC: A first-order Bayesian classifier. In *Proceedings of the 9th International Workshop on Inductive Logic Programming (ILP-99)*, pages 92–103. Springer, 1999.

Yoav Freund and Robert E. Schapire. Experiments with a new boosting algorithm. In *Proceedings of the 13th International Conference on Machine Learning*, pages 148–156, 1996.

Yoav Freund and Robert E. Shapire. Experiments with a new boosting algorithm. In *Machine Learning: Proceedings of the Thirteenth International Conference on Machine Learning*, 1996.

Johannes Fürnkranz, Dragan Gamberger, and Nada Lavrač. *Foundations of Rule Learning*. Springer, 2012.

Gemma C. Garriga, Petra Kralj, and Nada Lavrač. Closed sets for labeled data. *Journal of Machine Learning Research*, 9:559–580, 2008.

Ian Goodfellow, Jean Pouget-Abadie, Mehdi Mirza, Bing Xu, David Warde-Farley, Sherjil Ozair, Aaron Courville, and Yoshua Bengio. Generative adversarial nets. In *Advances in Neural Information Processing Systems*, pages 2672–2680, 2014.

Ian Goodfellow, Yoshua Bengio, Aaron Courville, and Yoshua Bengio. *Deep Learning*. The MIT Press, 2016.

Jiawei Han and Micheline Kamber. *Data Mining: Concepts and Techniques*. Morgan Kaufmann Publishers, 2001.

Trevor Hastie, Robert Tibshirani, and Jerome H. Friedman. *The Elements of Statistical Learning*. Springer-Verlag, 2001.

Jing He. Advances in data mining: History and future. In *Third International Symposium on Intelligent Information Technology Application (IITA 2009)*, volume 1, pages 634–636, 2009.

Sepp Hochreiter and Jürgen Schmidhuber. Long short-term memory. *Neural Computation*, 9(8): 1735–1780, 1997.

Kurt Hornik, Maxwell Stinchcombe, and Halbert White. Multilayer feedforward networks are universal approximators. *Neural Networks*, 2(5):359–366, 1989.

TaeHyun Hwang and Rui Kuang. A heterogeneous label propagation algorithm for disease gene discovery. In *Proceedings of SIAM International Conference on Data Mining*, pages 583–594, 2010.

Glen Jeh and Jennifer Widom. SimRank: A measure of structural-context similarity. In *Proceedings of the 8th ACM SIGKDD International Conference on Knowledge Discovery and Data Mining*, pages 538–543, 2002.

Ming Ji, Yizhou Sun, Marina Danilevsky, Jiawei Han, and Jing Gao. Graph regularized transductive classification on heterogeneous information networks. In *Proceedings of the 25th European Conference on Machine Learning and Principles and Practice of Knowledge Discovery in Databases*, pages 570–586, 2010.

Willi Klösgen. Explora: A multipattern and multistrategy discovery assistant. *Advances in Knowledge Discovery and Data Mining*, pages 249–271, 1996.

Thomas Kluyver, Benjamin Ragan-Kelley, Fernando Pérez, Brian Granger, Matthias Bussonnier, Jonathan Frederic, Kyle Kelley, Jessica Hamrick, Jason Grout, Sylvain Corlay, Paul Ivanov, Damián Avila, Safia Abdalla, Carol Willing, and Jupyter development team. Jupyter notebooks - a publishing format for reproducible computational workflows. In *Positioning and Power in Academic Publishing: Players, Agents and Agendas*, pages 87–90, 2016.

Risi Imre Kondor and John D. Lafferty. Diffusion kernels on graphs and other discrete input spaces. In *Proceedings of the 19th International Conference on Machine Learning*, pages 315–322, 2002.

Igor Kononenko and Matjaž Kukar. *Machine Learning and Data Mining: Introduction to Principles and Algorithms*. Horwood Publishing, 1st edition, 2007.

Petra Kralj Novak, Nada Lavrač, and Geoffrey I. Webb. Supervised descriptive rule discovery: A unifying survey of contrast set, emerging pattern and subgroup mining. *Journal of Machine Learning Research*, 10:377–403, February 2009.

Janez Kranjc, Roman Orač, Vid Podpečan, Nada Lavrač, and Marko Robnik-Šikonja. ClowdFlows: Online workflows for distributed big data mining. *Future Generation Computer Systems*, 68: 38–58, 2017.

Pat Langley. *Elements of Machine Learning*. Morgan Kaufmann, 1996.

Nada Lavrač and Anže Vavpetič. Relational and semantic data mining. In *Proceedings of the Thirteenth International Conference on Logic Programming and Nonmonotonic Reasoning*, pages 20–31, 2015.

Nada Lavrač and Sašo Džeroski. *Inductive Logic Programming: Techniques and Applications*. Ellis Horwood, 1994.

Yann LeCun, Léon Bottou, Yoshua Bengio, and Patrick Haffner. Gradient-based learning applied to document recognition. *Proceedings of the IEEE*, 86(11):2278–2324, 1998.

Yann LeCun, Yoshua Bengio, and Geoffrey Hinton. Deep learning. *Nature*, 521(7553):436, 2015.

Huan Liu and Hiroshi Motoda, editors. *Feature Extraction, Construction and Selection: a Data Mining Perspective*. Kluwer Academic Publishers, 1998.

Donna Maglott, Jim Ostell, Kim D. Pruitt, and Tatiana Tatusova. Entrez Gene: Gene-centered information at NCBI. *Nucleic Acids Research*, 33:D54–D58, 2005.

Ryszard S. Michalski. On the quasi-minimal solution of the covering problem. In *Proceedings of the 5th International Symposium on Information Processing (FCIP-69)*, volume A3 (Switching Circuits), pages 125–128, 1969.

Ryszard S. Michalski. Pattern recognition and rule-guided inference. *IEEE Transactions on Pattern Analysis and Machine Intelligence*, 2:349–361, 1980.

Ryszard S. Michalski, Jaime G. Carbonell, and Thomas M. Mitchell, editors. *Machine Learning: An Artificial Intelligence Approach, Vol. I*. Tioga, 1983.

Ryszard S. Michalski, Igor Mozetič, Jiarong Hong, and Nada Lavrač. The multi-purpose incremental learning system AQ15 and its testing application on three medical domains. In *Proceedings of the 5th National Conference on Artificial Intelligence*, pages 1041–1045, 1986.

Ingo Mierswa, Michael Wurst, Ralf Klinkenberg, Martin Scholz, and Timm Euler. YALE: Rapid prototyping for complex data mining tasks. In *Proceedings of the 12th ACM SIGKDD International Conference on Knowledge Discovery and Data Mining*, pages 935–940, 2006.

George A. Miller. WordNet: A lexical database for English. *Communications of the ACM*, 38(11): 39–41, November 1995.

Tom M. Mitchell. *Machine Learning*. McGraw Hill, 1997.

Stephen H. Muggleton, editor. *Inductive Logic Programming*. Academic Press, London, 1992.

Stephen H. Muggleton. Inverse entailment and Progol. *New Generation Computing*, 13(3–4): 245–286, 1995.

Kevin P. Murphy. *Machine Learning: A Probabilistic Perspective*. The MIT press, 2012.

Mark E. J. Newman. Clustering and preferential attachment in growing networks. *Physical Review E*, 64(2):025102, 2001a.

Mark E. J. Newman. The structure of scientific collaboration networks. *Proceedings of the National Academy of Sciences of the United States of America*, 98(2):404–409, 2001b.

Hiroyuki Ogata, Susumu Goto, Kazushige Sato, Wataru Fujibuchi, Hidemasa Bono, and Minoru Kanehisa. KEGG: Kyoto encyclopedia of genes and genomes. *Nucleic Acids Research*, 27(1): 29–34, 1999.

Lawrence Page, Sergey Brin, Rajeev Motwani, and Terry Winograd. The PageRank citation ranking: Bringing order to the web. Technical report, Stanford InfoLab, November 1999.

Judea Pearl. *Probabilistic Reasoning in Intelligent Systems: Networks of Plausible Inference*. Morgan Kaufmann, 1988.

Fabian Pedregosa, Gaël Varoquaux, Alexandre Gramfort, Vincent Michel, Bertrand Thirion, Olivier Grisel, Mathieu Blondel, Peter Prettenhofer, Ron Weiss, Vincent Dubourg, Jake Vanderplas, Alexandre Passos, David Cournapeau, Matthieu Brucher, Matthieu Perrot, Edouard Duchesnay, and Gilles Louppe. Scikit-learn: Machine learning in Python. *Journal of Machine Learning Research*, 12:2825–2830, 2011.

Gregory Piatetsky-Shapiro. Discovery, analysis, and presentation of strong rules. In *Knowledge Discovery in Databases*, pages 229–248. The MIT Press, 1991.

Michel Plantié and Michel Crampes. Survey on social community detection. In N. Ramzan et al., editor, *Social Media Retrieval*, pages 65–85. Springer, 2013.

J. Ross Quinlan. Discovering rules by induction from large collections of examples. In D. Michie, editor, *Expert Systems in the Micro Electronic Age*, pages 168–201. Edinburgh University Press, 1979.

J. Ross Quinlan. Induction of decision trees. *Machine Learning*, 1(1):81–106, 1986.

J. Ross Quinlan. Learning logical definitions from relations. *Machine Learning*, 5:239–266, 1990.

J. Ross Quinlan. Learning with continuous classes. In N. Adams and L. Sterling, editors, *Proceedings of the 5th Australian Joint Conference on Artificial Intelligence*, pages 343–348. World Scientific, 1992.

J. Ross Quinlan. *C4.5: Programs for Machine Learning*. Morgan Kaufmann, San Francisco, 1993.

Marko Robnik-Šikonja. Data generators for learning systems based on RBF networks. *IEEE Transactions on Neural Networks and Learning Systems*, 27(5):926–938, May 2016.

David E. Rumelhart and James L. McClelland, editors. *Parallel Distributed Processing: Explorations in the Microstructure of Cognition*, volume 1: Foundations. The MIT Press, Cambridge, MA, 1986.

David E. Rumelhart, Geoffrey E. Hinton, and Ronald J. Williams. Learning representations by back-propagating errors. *Nature*, 323(6088):533, 1986.

Gerard Salton, Andrew Wong, and Chungshu Yang. A vector space model for automatic indexing. *Communications of the ACM*, 18(11):613–620, 1975.

Robert E. Schapire, Yoav Freund, Peter Bartlett, and Wee Sun Lee. Boosting the margin: A new explanation for the effectiveness of voting methods. In Douglas H. Fisher, editor, *Machine Learning: Proceedings of the Fourteenth International Conference (ICML'97)*, pages 322–330. Morgan Kaufmann, 1997.

Bernhard Schölkopf and Alexander J. Smola. *Learning with Kernels: Support Vector Machines, Regularization, Optimization, and Beyond*. The MIT Press, 2001.

Ashwin Srinivasan. *The Aleph Manual*. University of Oxford, 2007. Online. Accessed 26 October 2020. URL: https://www.cs.ox.ac.uk/activities/programinduction/Aleph/.

Fabian M. Suchanek, Gjergji Kasneci, and Gerhard Weikum. YAGO: A core of semantic knowledge. In *Proceedings of the 16th International Conference on World Wide Web, WWW 2007, Banff, Alberta, Canada, May 8–12, 2007*, pages 697–706. ACM, 2007.

Yizhou Sun and Jiawei Han. *Mining Heterogeneous Information Networks: Principles and Methodologies*. Morgan & Claypool Publishers, 2012.

Yizhou Sun, Jiawei Han, Peixiang Zhao, Zhijun Yin, Hong Cheng, and Tianyi Wu. RankClus: Integrating clustering with ranking for heterogeneous information network analysis. In

Proceedings of the International Conference on Extending Data Base Technology, pages 565–576, 2009a.

Yizhou Sun, Yintao Yu, and Jiawei Han. Ranking-based clustering of heterogeneous information networks with star network schema. In *Proceedings of the 15th ACM SIGKDD International Conference on Knowledge Discovery and Data Mining*, pages 797–806, 2009b.

Oron Vanunu, Oded Magger, Eytan Ruppin, Tomer Shlomi, and Roded Sharan. Associating genes and protein complexes with disease via network propagation. *PLoS Computational Biology*, 6 (1), 2010.

Vladimir N. Vapnik. *The Nature of Statistical Learning Theory*. Springer, 1995.

Ashish Vaswani, Noam Shazeer, Niki Parmar, Jakob Uszkoreit, Llion Jones, Aidan N Gomez, Łukasz Kaiser, and Illia Polosukhin. Attention is all you need. In *Advances in Neural Information Processing Systems*, pages 5998–6008, 2017.

Willem Waegeman, Krzysztof Dembczyński, and Eyke Hüllermeier. Multi-target prediction: A unifying view on problems and methods. *Data Mining and Knowledge Discovery*, 33(2):293–324, 2019.

Ian H. Witten, Eibe Frank, and Mark A. Hall. *Data Mining: Practical Machine Learning Tools and Techniques*. Morgan Kaufmann, 3rd edition, 2011.

David H. Wolpert. Stacked generalization. *Neural Networks*, 5(2):241–260, 1992.

Stefan Wrobel. An algorithm for multi-relational discovery of subgroups. In *Proceedings of the 1st European Symposium on Principles of Data Mining and Knowledge Discovery (PKDD-97)*, pages 78–87, 1997.

Bo Yang, Dayou Liu, and Jiming Liu. Discovering communities from social networks: Methodologies and applications. In *Handbook of Social Network Technologies and Applications*, pages 331–346. Springer, 2010.

Wayne W. Zachary. An information flow model for conflict and fission in small groups. *Journal of Anthropological Research*, 33:452–473, 1977.

Mohammed J. Zaki and Ching-Jui Hsiao. Efficient algorithms for mining closed itemsets and their lattice structure. *IEEE Transactions on Knowledge and Data Engineering*, 17(4):462–478, April 2005.

Dengyong Zhou, Olivier Bousquet, Thomas Navin Lal, Jason Weston, and Bernhard Schölkopf. Learning with local and global consistency. *Advances in Neural Information Processing Systems*, 16(16):321–328, 2004.

Zhi-Hua Zhou and Ji Feng. Deep forest: Towards an alternative to deep neural networks. In *Proceedings of the Twenty-Sixth International Joint Conference on Artificial Intelligence, IJCAI-2017*, pages 3553–3559, 2017.

Chapter 3
Text Embeddings

Text embeddings are a subfield of data representation, studying different numerical text representations, from sparse to dense, and different embedding techniques, from matrix factorization to deep neural approaches. This chapter starts by presenting two technologies, transfer learning and language models, in Sect. 3.1, used as a necessary background for understanding advanced embedding technologies explained later in this chapter. The rest of this chapter follows the historical development of text embeddings, starting with embeddings based on word cooccurrence statistics in Sect. 3.2, and continuing with neural network-based embeddings in Sect. 3.3, where we first present word2vec and GloVe, followed by contextual embeddings, such as ELMo and BERT embeddings, based on deep neural language models. In Sect. 3.4, we discuss embeddings of larger text units, such as sentences and documents. Cross-lingual embeddings are discussed in Sect. 3.5. Section 3.6 addresses the problem of quality assessment of embedding methods. While embeddings are usually assessed with extrinsic evaluation on a selected downstream text analysis task (e.g., sentiment analysis, text classification, or information retrieval), several intrinsic evaluation approaches exist, including the analogy task that we briefly present. The chapter concludes by presenting selected methods implemented in Jupyter Python notebooks in Sect. 3.7.

3.1 Background Technologies

The two technologies, presented in this section, are important for understanding modern embeddings methods, presented later in this chapter. The concepts of transfer learning is outlined in Sect. 3.1.1 and language models are defined in Sect. 3.1.2.

© Springer Nature Switzerland AG 2021
N. Lavrač et al., *Representation Learning*,
https://doi.org/10.1007/978-3-030-68817-2_3

3.1.1   Transfer Learning

The idea of transfer learning is to solve a selected problem and transfer the gained knowledge to solve other related problems. For example, knowledge of language gained when learning to predict the next word in a sequence can be used when classifying the sentiment of texts; or knowledge obtained from mobile sensors when predicting the users' behavior can be used to predict mobility disorders typical for Parkinson's disease. The reuse of information from previously learned tasks is a significant asset for a broader application of machine learning and can lower dataset annotation cost.

While transfer learning is by no means limited to deep neural networks (DNNs), this topic has spurred a strong interest in deep learning community. In DNNs, transfer learning is applied in two ways: (1) as reuse and retraining of a network trained on a related task, and (2) as training of several related tasks together.

In **model reuse** a model developed for one task is reused as the starting point for a second task model (Pratt et al. 1991). For many imaging and text related problems, pre-trained models are available and can be used as the starting point. This may greatly reduce the resources required to develop neural network models on these problems.

Multi-task learning (Caruana 1997) tries to train models on several related tasks together. During training, the related tasks share representation (e.g., neurons in a deep neural network). This reduces the likelihood of overfitting as related tasks serve as a regularization mechanism that enforces better generalization.

3.1.2   Language Models

Language model (LMs) are probabilistic prediction models that learn distributions of words in a given language (Manning and Schütze 1999). For example, for a sequence of words $w_1 w_2 \ldots w_n$, the language model computes the probability $p(w_1 w_2 \ldots w_n)$ of observing the sequence in the given corpus. Frequently, language models predict the next word in a sequence by computing the conditional probability $p(w_{n+1} | w_1 w_2 \ldots w_n)$ that the word w_{n+1} follows the word sequence $w_1 w_2 \ldots w_n$. Traditionally, language models were trained on large text corpora using n-grams, assuming the independence of words beyond a certain distance. For example, if we assume that a word only depends on the previous one in the sequence, i.e. $p(w_{n+1} | w_1 w_2 \ldots w_n) = p(w_{n+1} | w_n)$, we can only count all joint occurrences of pairs of words. Let $c(w_i)$ denote the number of appearances of the word w_i in the corpus, and $c(w_i w_j)$ the number of joint occurrences of words w_i and w_j. To estimate the probability of observing the next word w_{n+1} in the sequence using bigrams we compute:

$$p(w_{n+1}|w_n) = \frac{c(w_n w_{n+1})}{c(w_n)}.$$

The probability of a sequence of words is retrieved using the Bayes chain rule for conditional probabilities:

$$p(w_1 w_2 \ldots w_n) = p(w_1) \cdot p(w_2|w_1) \ldots p(w_n|w_1 w_2 \ldots w_{n-1}).$$

Using the Markov independence assumption for sequences longer than two words, we get:

$$p(w_1 w_2 \ldots w_n) = p(w_1) \cdot p(w_2|w_1) \ldots p(w_n|w_{n-1}).$$

The remaining conditional probabilities in the above expression can be estimated using bigram and unigram counts.

In real texts, the Markov independence assumption is not true even for very long sequences (think of split verbs or dependent sentences separating a noun and its verb). The n-gram counting method, therefore, does not work well. Additionally, frequencies of n-grams for $n > 3$ become statistically very unreliable even with huge corpora.

Lately, language models are trained using deep neural networks. If a neural network is trained to predict the next word in a sequence from a large text corpus, the sequences can be much longer and we still get reliable results. Language models can also be trained in the reverse direction, i.e. for a backwards language model ($\overleftarrow{LM}$) we train a network to predict the probability $p(w_i|w_{i+1} w_{i+2} \ldots w_{i+k})$ that the word w_i appears before the sequence $w_{i+1} w_{i+2} \ldots w_{i+k}$. Further generalization of LMs are referred to as *masked language models* (Devlin et al. 2019), which predict a missing word anywhere in a sequence, mimicking a gap filling test

$$p(w_i|w_{i-b} w_{i-b+1} \ldots w_{i-1} w_{i+1} \ldots w_{i+f}).$$

For the n-gram approach this would be unfeasible but neural networks are flexible enough and can successfully predict the gaps.

3.2 Word Cooccurrence-Based Embeddings

To process natural language with modern machine learning tools, texts (words, phrases, sentences, and documents) must be transformed into vectors of real numbers. While first approaches represented each word in a high dimensional space with one coordinate per word (so-called one-hot encoding and bag-of-words representation), nowadays, word embeddings refer to word representations in a numeric vector space of a much lower dimensionality. The methods for producing word embeddings rely on the distributional hypothesis of Harris (1954), which states that words occurring in similar contexts tend to have similar meanings. Exploiting word cooccurrence properties leads to word vectors that reflect semantic similarities and dissimilarities: similar words are close in the embedding space, and conversely, in the embedding space, dissimilar words are far from each other. These representations can be used to assess text similarity by computing distances between their vectors. This is applicable in many text processing tasks, such as information retrieval, document classification, topic detection, question answering, sentiment analysis, machine translation, etc.

Word representations can be divided into two groups: sparse word cooccurrence-based embeddings and dense word cooccurrence-based embeddings.

Sparse word embeddings.   These embeddings, presented in Sect. 3.2.1, cover one-hot encoding and bag-of-words (BoW) representations. Various weighting schemes and similarity measures used in BoW representations are presented in Sects. 3.2.2 and 3.2.3, respectively. Sparse matrix representation of documents, using a term frequency weighting scheme, is illustrated in Sect. 3.2.4.

Dense word embeddings.   These embeddings include the Latent Semantic Analysis (LSA) approach characterized by reducing the dimensionality of the word cooccurrence matrix with the Singular Value Decomposition (SVD) approach presented in Sect. 3.2.5, as well as the Probabilistic LSA (PLSA) and Latent Dirichlet Analysis (LDA) topic-based embedding approaches presented in Sect. 3.2.6.

Note that modern dense embedding approaches use neural networks to produce embeddings. The main representatives of this group are word2vec, GloVe, ELMo, and BERT, presented in Sect. 3.3.

3.2.1 Sparse Word Cooccurrence-Based Embeddings

One-hot word encoding.   The simplest representation of a word is a so-called one-hot vector, which has the dimension of the dictionary and contains 1 in the coordinate representing the given word and 0 in all other coordinates. As an example, let us take the sentence:

Tina lives in a house.

In preprocessing (see Sect. 2.2), after removing the so-called stop words, three words remain: *Tina lives house*. Using one-hot vectorization and a realistic dictionary (e.g., 50,000 words appearing in a large document corpus), the three words would be represented with one-hot vectors shown in Fig. 3.1. We can quickly see several shortcomings of this representation. For example, two similar words like *house* and *cottage* are assigned two different coordinates; their distance in this representation is, therefore, the same as the distance between two arbitrary unrelated words such as *house* and *Tina*. Disregarding word similarity is a severe weakness when addressing typical language processing tasks, such as information retrieval, semantic search, or classification.

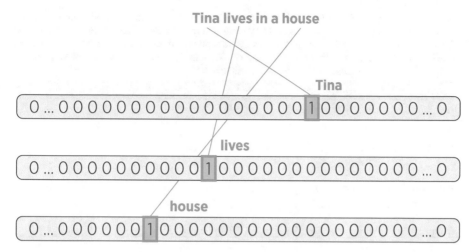

Fig. 3.1 Example of one-hot vector representation of words. In this example, words are not lemmatized

Bag-of-words. While one-hot encoding represents a single word, sentences can be represented by placing 1 in all vector coordinates corresponding to the words of the sentence. Similarly, a representation of a document would contain non-zero entries for all the words in the document. As more occurrences of the same word mean that the word is characteristic of the document, the values in document vectors typically denote the number of occurrences for each word. This representation of a document is referred to as bag-of-words (BoW) (Jones 1972; Feldman and Sanger 2006). A simple example of representing two sentences, *Tina lives in a house* and *The house is large*, after removing the stop words (in the context of a corpus of 1000 documents with a dictionary of 50,000 words), is shown in Table 3.1. For large vocabularies, the BoW representation of documents is problematic due to large sparse vectors and the assumption of independent coordinates. These issues are resolved with dense vector embeddings described in Sects. 3.2.5 and 3.3.

3.2.2 Weighting Schemes

In different text analysis tasks, the individual features (words or terms) in the n-dimensional BoW vector representation shall not be equally important. To address this issue, Weighting schemes assign importance to features. The four most popular weighting schemes are listed below.

Binary. In this simplest weighting model, a feature weight is 1 if the corresponding feature (word or term) is present in the document; otherwise, it is zero.

Table 3.1 Example BoW representation of a corpus of 1000 documents with a dictionary of 50,000 words. The two shown documents each containing a single sentence. *Tina lives in a house* and *The house is large*. In the BoW representation, rows represent individual documents, and columns represent individual words. Entry $x_{i,j}$ at position (i, j) indicates the appearance of word w_j in document d_i. In this example, words are not lemmatized

Document/Word	w_1	...	*house*	...	*large*	...	*lives*	...	Tina	...	w_{50000}
d_1	0	...	1	...	0	...	0	...	0	...	0
$\vdots$		$\ddots$	$\vdots$	$\ddots$	$\vdots$	$\ddots$	$\vdots$	$\ddots$	$\vdots$	$\ddots$	$\vdots$
Tina lives in a house.	0	...	1	...	0	...	1	...	1	...	0
$\vdots$		$\ddots$	$\vdots$	$\ddots$	$\vdots$	$\ddots$	$\vdots$	$\ddots$	$\vdots$	$\ddots$	$\vdots$
The house is large.	0	...	1	...	1	...	0	...	0	...	0
$\vdots$		$\ddots$	$\vdots$	$\ddots$	$\vdots$	$\ddots$	$\vdots$	$\ddots$	$\vdots$	$\ddots$	$\vdots$
d_{1000}	0	...	0	...	0	...	0	...	0	...	1

Term occurrence.   In this model, a feature weight is equal to the number of its occurrences. This weight is often better than a simple binary value since frequently occurring features are more relevant to the given text.

Term frequency (TF).   In this model, the weight is derived from the term occurrence by dividing the vector components by the sum of all vector's weights. The reasoning is similar to the term occurrence, with normalization with respect to the document size.

Term Frequency-Inverse Document Frequency (TF-IDF).   In this scheme, the weight of a feature (word or term) does not depend only on the document but also on the corpus to which the document belongs. The weight increases with the number of its occurrences in the document (TF), as this indicates its relevance for the given document. This effect is counter-balanced by the inverse of its frequency (IDF) in the document corpus, as a high frequency in all corpus documents indicates a commonly appearing word (such as an article) and not necessarily the word's relevance (Jones 1972; Salton and Buckley 1988). TF-IDF is defined as:

$$\text{TF} - \text{IDF}(w_i, d_j) = \text{TF}(w_i, d_j) \cdot \log \frac{|D|}{\text{DF}(w_i)}, \tag{3.1}$$

where $\text{TF}(w_i, d_j)$ is the frequency of the given word (or term) w_i inside document d_j, $|D|$ is the number of all documents in document corpus D, and $\text{DF}(w_i)$ is the number of documents that contain w_i. The reasoning behind the TF-IDF weighting is to lower weights of words that appear in many documents, as this is usually an indication of them being less important.

In addition to the above most frequently used feature weighting schemes, there are numerous other schemes available. The Okapi BM25 weighting scheme (Robertson and Walker 1994) is e.g., used to evaluate how well a query of several terms matches a given document. The χ^2 weighting scheme (Debole and Sebastiani 2004), which is defined for documents with assigned class labels, attempts to correct a drawback of the TF scheme (which is not addressed by the TF-IDF scheme) by considering the class value of processed documents. The scheme penalizes terms that appear in documents of all classes and favors the terms that are specific to individual classes. The information gain (IG) scheme (Debole and Sebastiani 2004) uses class labels to improve term weights. It applies an information-theoretic approach by measuring the amount of information about one random variable (the class of a document) gained by knowledge of another random variable (the appearance of a given term). The gain ratio (GR) scheme is similar to the information gain but is normalized by the total entropy of the class labels in the data set. The Δ-idf (Delta-IDF) (Martineau and Finin 2009) and relevance frequency (RF) schemes (Lan et al. 2009) attempt to merge ideas of TF-IDF and both above class-based schemes by penalizing both common terms as well as the terms that are not class-informative.

3.2.3 Similarity Measures

The most common distance measures are the *dot product* and the *cosine similarity*.

Dot product. Dot product $x \cdot y$ of vectors x and y is the sum of the products of feature values of vectors.

Cosine similarity. This measure is computed as the cosine of angle α between the two non-zero vectors x and y (which equals the inner product of the same vectors normalized to both have length 1). The *cosine similarity* is hence defined as follows:

$$\text{similarity}(x, y) = \cos(\alpha) = \frac{x \cdot y}{||x|| \cdot ||y||} = \frac{\sum_1^n x_i \cdot y_i}{\sqrt{\sum_1^n x_i^2} \cdot \sqrt{\sum_1^n y_i^2}}, \quad (3.2)$$

where x_i and y_i are components of vectors x and y, respectively. The cosine similarity measure can be used to find similar documents, or documents similar to a given query provided that the query is also represented as a vector.

If the two vectors have been normalized before their distance computation, the dot product and cosine similarity produce identical results. Otherwise, the choice of the similarity measure is related to the choice of the weighting model.

A well-tested combination for computing the similarity of documents in the feature space is the TF-IDF weighting with the cosine similarity. We can write the weighted version of the cosine similarity between two documents, x and y, using the vocabulary V, as follows:

$$\text{similarity}(x, y) = \frac{\sum_{i=1}^{|V|} (\text{TF-IDF}(w_i, x) \cdot \text{TF-IDF}(w_i, y))}{\sqrt{\sum_{i=1}^{|V|} \text{TF-IDF}(w_i, x)^2} \cdot \sqrt{\sum_{i=1}^{|V|} \text{TF-IDF}(w_i, y)^2}}. \quad (3.3)$$

Table 3.2 An example of a term-document matrix, with terms (words) in rows and documents in columns. Entry $x_{i,j}$ at position (i, j) indicates the frequency of term w_i in document d_j

Word/Document	d_1	d_2	...	d_j	...	d_m
w_1	0	0	...	8	...	0
w_2	1	0	...	0	...	7
⋮	⋮	⋮	⋱	⋮	⋱	⋮
w_i	0	5	...	0	...	1
⋮	⋮	⋮	⋱	⋮	⋱	⋮
w_n	0	0	...	3	...	0

Table 3.3 Schema of a term-term matrix, with words (terms) in rows and words (contexts) in columns. Entry $x_{i,j}$ at position (i, j) indicates the frequency of term w_i in the context of term w_j

Word/Word	w_1	w_2	...	w_j	...	w_n
w_1	0	0	...	0	...	6
w_2	3	0	...	0	...	0
⋮	⋮	⋮	⋱	⋮	⋱	⋮
w_i	17	0	...	0	...	0
⋮	⋮	⋮	⋱	⋮	⋱	⋮
w_n	0	0	...	21	...	0

3.2.4   Sparse Matrix Representations of Texts

This section presents sparse matrix representations of texts through examples using a term frequency weighting scheme.

In the BoW representation, a document representation is composed of word representations. For a collection of documents, we can merge their representations column-wise as illustrated in Table 3.2. The resulting matrix is referred to as *term-document matrix*. One row of this matrix represents one word (term) of the vocabulary. The matrix contains non-zero entries for the documents where the given word appears. We assume that there are n words in the vocabulary and m documents in the corpus. Some text mining tasks use the term-document matrix transpose and work with the document-term matrix (documents in rows).

Instead of a whole document, we can use shorter text units as a context for each word, for example, a section, a paragraph, a sentence, or a few preceding and consecutive words. For these contexts, we can observe the cooccurrence of words and store the results in a matrix, referred to as term-term matrix, illustrated in Table 3.3. In this matrix, both rows and columns represent words, and the entry $x_{i,j}$ contains the number of times term w_i is located in the same context as term w_j.

This term-term matrix computationally supports the distributional semantics, summarized by the famous observation of Firth (1957): "You shall know a word by the company it keeps!". If the term-term matrix is based on a large corpus of documents, it contains very relevant information about the similarity of words: which words frequently appear together (e.g., food, restaurant, menu), and which words appear within similar contexts (e.g., drive, transport, move).

Unfortunately, the term-term matrix is very sparse, which is not surprising. If we take a realistic (but not very large) vocabulary of 100,000 words, the corpus from where we would take contexts would need to have at least 10 billion words for each word to have at least one occurrence with every other word. In reality, more than 99% of entries in typical term-term matrices are zeros.

3.2.5 Dense Term-Matrix Based Word Embeddings

This section presents matrix-factorization based dense word embeddings, while dense neural word embeddings are presented in Sect. 3.3. In this group of embedding approaches, the construction of word embeddings is based on the idea from linear algebra to compress the information encoded in a sparse matrix representation. This compression is done with the *Singular-Value Decomposition* (SVD) technique in an approach named *Latent Semantic Analysis* (LSA).

Singular-value decomposition. Singular-value decomposition of a $v \times c$ dimensional real matrix X is a factorization of the form $X = W \cdot \Sigma \cdot C$ with the following components.

- W is a $v \times m$ dimensional orthogonal matrix, i.e. $W \cdot W^T = I_v$, where W^T is the transpose of W, and I_v is the identical matrix.
- Σ is a $m \times m$ dimensional rectangular diagonal matrix with non-negative real numbers on the diagonal. The diagonal entries σ_i of Σ are referred to as singular values of X. The number of non-zero singular values m is equal to the rank of matrix X.
- C is a $m \times c$ dimensional real orthogonal matrix, i.e. $C \cdot C^T = I_m$.

The columns of W are referred to as left-singular vectors of $X \cdot X^T$, while the rows of C are right-singular vectors, and are the eigenvectors of $X^T \cdot X$. The non-zero singular values of X (i.e. the diagonal entries of Σ) are the square roots of the non-zero eigenvalues of $X^T \cdot X$ and $X \cdot X^T$. The schema for this decomposition is shown in Fig. 3.2. The diagonal entries σ_{ii} are usually ordered in the descending order.

Latent Semantic Analysis. If we apply singular-value decomposition to a term-document matrix X, the procedure is named Latent Semantic Analysis (Deerwester et al. 1990). An outline of the method is presented in Fig. 3.3.

- In the first step, matrix factorization with SVD decomposes matrix X into orthogonal word matrix W, singular values matrix Σ, and orthogonal context matrix C. The columns of W are linear combinations of the original words and represent the base vectors. The entire matrix W is dense and contains dense embeddings of words. Each row represents one word from the vocabulary.
- If we set the smallest elements of Σ to zero (keeping, e.g., 300 largest elements), we effectively truncate also the matrices W and C (retaining the

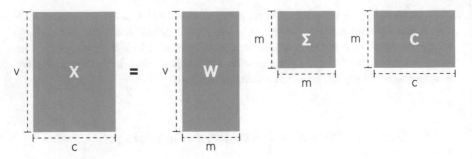

Fig. 3.2 Singular value decomposition of matrix X

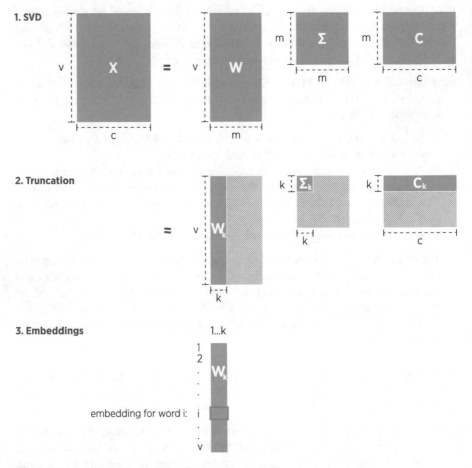

Fig. 3.3 Latent Semantic Analysis of term-term matrix X

first 300 columns in W and 300 rows in C), as illustrated in Step 2 of Fig. 3.3. The product of these truncated W, Σ, and C is an approximation of the original matrix X.

- The remaining matrix W_k contains the embeddings of words, i.e. the embedding of word i is the vector contained in the i-th row of W_k, as shown in Step 3 of Fig. 3.3. The dimensionality of this vector is k and typically ranges from 100 to 500.

The problem of matrix factorization based embedding methods is a too strong impact of frequent words. Several methods try to remedy that. For example, Bullinaria and Levy (2007) weight entries of the term-term matrix with positive point-wise mutual information.

3.2.6 Dense Topic-Based Embeddings

Note that LSA can also be viewed from the perspective of topic modeling, which assumes that documents are composed of some underlying topics. The eigenvectors produced by SVD decomposition can be seen as latent concepts, as they characterize the data by relating words used in similar contexts. Unfortunately, the coordinates in the new space do not have an obvious interpretation.

Probabilistic Latent Semantic Analysis. In PLSA (Hofmann 1999), each document can be viewed as a mixture of various topics. Rather than applying SVD to derive latent vectors by factorization, PLSA fits a statistical latent class model on a word-document matrix using tempered expectation maximization. This process generates a low-dimensional latent semantic space in which coordinates are topics represented with probability distributions over words, i.e. sets of words with a varying degree of membership to the topic. In this representation, we can see sets of words as latent concepts, and documents are probabilistic mixtures of these topics. The number of dimensions in the new space is determined according to the statistical theory for model selection and complexity control.

Latent Dirichlet Allocation. A shortcoming of PLSA is that it has to estimate probabilities from topic-segmented training data. This is addressed by the popular Latent Dirichlet Allocation (Blei et al. 2003), which takes a similar latent-variable approach but assumes that the topic distribution has a sparse Dirichlet prior. LDA supports the intuition that each document covers only a small set of topics and that in each topic, only a small set of words appear frequently. These assumptions enable better disambiguation of words and a more precise assignment of documents to topics. Still, the topics that LDA returns are not semantically strongly defined, as they rely on the likelihood of term cooccurrence. A word may occur in several topics with different probabilities, but for each topic in which it appears it will have a different set of cooccurring words. Note that LDA is capable of producing vector representations of both words and documents.

3.3   Neural Word Embeddings

As deep neural networks became the predominant learning method for text analysis, they also gradually became the method of choice for text embeddings. A procedure common to these embeddings is to train a neural network on one or more semantic text classification tasks and then take the weights of the trained neural network as a representation for each text unit (word, n-gram, sentence, or document). The labels required for training such a classifier originate from huge corpora of texts. Typically, they reflect word cooccurrence. Examples are predicting the next and previous word in a sequence or filling in missing words (also named the *cloze test*). Representation learning can be extended with other related tasks, such as prediction if two sentences are sequential (named sentence entailment). The positive instances for learning are obtained from the text in the given corpus. The negative instances are mostly obtained with negative sampling (sampling from instances that are unlikely to be related).

In Sect. 3.3.1, we present the word2vec method that trains a shallow neural network and produces a single vector for each word; for example, word *bank*, denoting a financial institution, and word *bank*, denoting a land alongside a river, produce the same vectors. In Sect. 3.3.2, we introduce the GloVe embeddings, which adapts the ideas from statistical word cooccurrence (i.e. LSA) to the neural network setting. The most advanced text representations, which we present in Sect. 3.3.3, are contextual word embeddings, which train deep networks and combine weights from different layers to produce unique vectors for each word occurrence, based on the context (typically sentence) it appears in.

3.3.1   Word2vec Embeddings

Mikolov et al. (2013) proposed a word2vec word embedding method that uses a single layer feed-forward neural network to *predict* word cooccurrence. The prediction model was trained on a huge Google News data set, consisting of documents with about 100 billion words. Once trained, the weights of the hidden layer in this network were used as word embeddings. The pretrained 300-dimensional vectors for 3 million English words and phrases are publicly available.[1] Word2vec consists of two related methods: the Continuous Bag-of-Words (CBOW) and the skip-gram method. Both methods construct a neural network to classify cooccurring words by taking a word and its d preceding and succeeding words, e.g., ± 5 words.

- The CBOW method takes the neighboring words and predicts the central word.
- The skip-gram variant takes the word and predicts its neighborhood.

---

[1] https://code.google.com/archive/p/word2vec/.

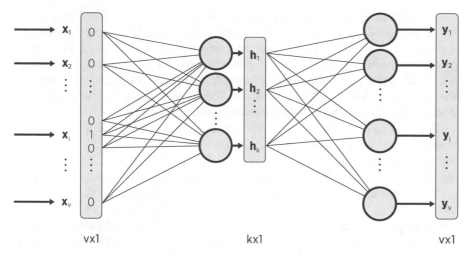

Fig. 3.4 Schema of the word2vec word embedding. The input and output of the network are one-hot encoded vectors of the dimension v (i.e. $v \times 1$), and the hidden layer, which determines the dimensionality of the embeddings is a vector of dimension k (i.e. $k \times 1$)

Note that the actual neural network architecture is similar in both cases, i.e. there is one word on the input (either the neighboring word for CBOW or the central word for the skip-gram method) and one word on the output, both represented in one-hot encoding as shown in Fig. 3.4. Empirical evaluations have shown a slight advantage of the skip-gram model over CBOW for many tasks; therefore, we focus on it in the rest of this section.

The words and their contexts (one word at a time) appearing in the training corpus constitute the training instances of the classification problem. Let us assume that we have the context window of size $d = 2$ and the sentence:

Tina is watching her linear algebra lecture with new glasses.

For the word *linear*, the following positive instances are generated:

(linear, watching)
(linear, her)
(linear, algebra)
(linear, lecture)

The first word of the training pair is presented at the input of the network in the one-hot-encoding representation, illustrated in Fig. 3.4. The network is trained to predict the second word. The difference in prediction is evaluated using a loss function. For a sequence of T training words $w_1, w_2, w_3, \ldots, w_T$, the skip-gram model maximizes the average log probability:

$$\frac{1}{T} \sum_{t=1}^{T} \sum_{-d \leq j \leq d, j \neq 0} \log p(w_{t+j}|w_t).$$

To make the process efficient for the 100 billion Google News corpus, the actual implementation uses several approximation tricks. The biggest problem is the estimation of $p(w_{t+j}|w_t)$, which normally requires a computation of the dot product between w_t and all other words in the vocabulary. Mikolov et al. (2013) solved this issue with negative sampling, which replaces $\log p(w_{t+j}|w_t)$ terms in the objective function with the results of logistic regression classifiers trained to distinguish between similar and dissimilar words.

Once the network is trained with word2vec, vectors for each word in the vocabulary can be generated. As Fig. 3.4 shows, one-hot encoding of the input word only activates one input connection for each hidden layer neuron. The weights on these connections constitute the embedding vector for the given input word.

The properties of the resulting word embeddings depend on the size of the context. For a small number of neighboring words (e.g., ± 5 words), we get embeddings that perform better on syntactic tasks. For larger neighborhoods (e.g., ± 10 words), the embeddings better express semantic properties. There is also some difference between the CBOW and the skip-gram variant. Both CBOW and skip-gram models perform comparably for syntactic problems, but the skip-gram variant works better for semantic problems.

3.3.2 GloVe Embeddings

LSA and related statistical methods, outlined in Sect. 3.2.5, take the whole term-document or term-term matrix into account and efficiently leverage its statistical information. However, experiments show that these methods work relatively poorly on the word analogy task (see its description in Sect. 3.6), indicating a sub-optimal vector space structure. A more advanced method exploiting the statistics stored in the term-term matrix, named GloVe (Global Vectors), was proposed by Pennington et al. (2014). This method presents training of word embeddings as an optimization problem. It uses stochastic optimization on word cooccurrences, using a context window 10 words before and after a given word, but decreasing the impact of more distant words. Instead of using the count of cooccurrence of word w_i with word w_j within a given context, or its entropy-based transformation, the GloVe method operates on a ratio of cooccurrence of words w_i and w_j, relatively to a context word w_t. Let P_{it} denote the probability of seeing words w_i and w_t together, which is computed by dividing the number of times w_i and w_t appeared together (c_{it}) by the total number of times word w_i appeared in the corpus (c_i). Similar notation is used for P_{jt}. For context words w_t, related to word w_i but not to word w_j, the ratio of probabilities $\frac{P_{it}}{P_{jt}}$ is large. Inversely, for words w_t related to word w_j but not to word w_i, this ratio is small. For words w_t, related to both w_i or w_j, or unrelated to both, the ratio is close to 1. Thus, this ratio can distinguish relevant related words from irrelevant ones. It also discriminates well between different degree of relevance for relevant words. For a given vocabulary V, GloVe minimizes the following loss function:

$$Loss_{\text{GloVe}} = \sum_{i=1}^{|V|} \sum_{j=1}^{|V|} f(C_{ij})(\hat{w}_i{}^T \hat{w}_j + b_i + b_j - \log C_{ij})^2,$$

where C_{ij} is a cooccurrence count of word vectors $\hat{w}_i, \hat{w}_j \in \mathbb{R}^k$ in the contexts of neighboring words observed in the corpus (by default, the dimensionality of embedding vectors $k = 100$). Function $f(C_{ij})$ is a weighting function that assigns relatively lower weight to rare and higher weight to frequent cooccurrences, and b_i and b_j are biases corresponding to words w_i and w_j.

GloVe is comparable and sometimes better than the popular word2vec (described in Sect. 3.3.1). The shortcoming of both methods is that they produce a single vector for each word, disregarding its different meanings (e.g., the word *plant* has a single vector, disregarding that it can represent a living organism or an industrial unit). This deficiency is addressed by contextual word embeddings, described in Sect. 3.3.3.

3.3.3 Contextual Word Embeddings

With the development of word2vec and GloVe, described in Sects. 3.3.1 and 3.3.2, the text mining community gained powerful tools. In particular, word2vec precomputed embeddings soon became a default choice for the first layers of many classification deep neural networks. The problem with word2vec embeddings is their failure to express polysemous words. During its training, all senses of a given word (e.g., *paper* as a material, as a newspaper, as a scientific work, and as an exam) contribute relevant neighboring words in proportion to their frequency in the training corpus. This causes the final vector to be placed somewhere in the weighted middle of all words' meanings. Consequently, rare meanings of words are poorly expressed with word2vec, and the resulting vectors do not offer good semantic representations. For example, none of the 50 closest vectors of the word *paper* is related to science.

The idea of contextual word embeddings is to generate a different vector for each context a word appears in. The context is typically defined sentence-wise. This solves the problems with word polysemy to a large extent. The context of a sentence is mostly enough to disambiguate different meanings of a word for humans; the same is mostly true for the learning algorithms. This section describes different ideas on how to take word context into account, used in modern embeddings. There are two baseline technologies common to them, the concepts of transfer learning and language models, which were presented in Sect. 3.1. Considering these technologies will allow the reader to understand the currently most successful approaches to contextual word embeddings, ELMo, ULMFit, BERT, and UniLM, presented in this section.

As described in Sect. 3.3.1, the word2vec method uses neural networks to build language representation by predicting the word based on its nearby words (i.e. the context). As a result of learning, each word is represented with a single vector. The neural network training takes pairs of neighboring words and uses one as an input and the other as the prediction. This procedure resembles a language model, but in a very rudimentary form: there are only two words involved in learning, the position of each word in a sequence is unknown, and so is the distance between the two words.

Several researchers noticed similar training procedures for word embeddings and *language models* (see Sect. 3.1.2) when used to train the embeddings. These approaches combine several tasks, like forward and backward language model, and train a neural network simultaneously on all of them. It is no longer trivial to extract vector representations from these models; the resulting deep neural networks store the knowledge about the language in weights spread over several layers. A frequently used approach is to concatenate weights from several layers into a vector, but often it is more convenient to use the whole pretrained language model as a starting point and refine its weights further during training on a specific task of interest. Such an approach corresponds to *transfer learning* (see Sect. 3.1.1), which is often used for modern word embeddings, indicating that the knowledge about the language structure is transferred from the language model to a downstream task. Below we outline the ideas of the main representatives of contextual embeddings: ELMo, ULMFit, BERT, and UniLM.

ELMo (Embeddings from Language Models).   The ELMo embeddings model, developed by Peters et al. (2018), is an example of a large pretrained transfer learning model. The actual embeddings are constructed from the internal states of a bidirectional LSTM neural network (see Sect. 2.1.7, on page 29). Higher-level layers capture context-dependent aspects of the input, while lower-level layers capture aspects of the syntax. ELMo treats its input on a character-level and builds words from subword tokens (groups of characters). To train the ELMo network, we put one sentence at a time on the input. After being processed by several layers, the representation of each word depends on the whole sentence. In this way, it reflects the contextual features of the input text and thereby the polysemy of words. For an explicit word representation, one can use only the top layer. Still, more frequently, practitioners combine all the layers into a vector. The representation of a token t_k at position k is composed from:

$$R_k = \{x_k^{LM}, \overrightarrow{h}_{k,j}^{LM}, \overleftarrow{h}_{k,j}^{LM} \mid j = 1, \ldots, L\}, \tag{3.4}$$

where L is the number of LSTM layers (ELMo uses $L = 2$), index j refers to the level of the bidirectional LSTM network, x is the initial token representation obtained via convolution of characters, and h^{LM} denotes hidden layers of the forward or backward language model LM.

It was shown that ELMo outperforms previous pretrained word embeddings like word2vec and GloVe on many Natural Language Processing (NLP) tasks, e.g., question answering, named entity extraction, sentiment analysis, textual entailment, semantic role labeling, and coreference resolution. If explicit vectors are required, ELMo's compact three-layer architecture may be preferable to BERT's deeper architecture.

ULMFiT (Universal Language Model Fine-Tuning). This model, developed by Howard and Ruder (2018), is based on transfer learning, which aims to transfer knowledge captured in a source task to one or more target tasks. ULMFiT uses language models as the source task. Howard and Ruder (2018) claim that language models (LMs) are the ideal source task for language problems as they capture many facets of language relevant for downstream tasks, e.g., long-term dependencies, hierarchical relations, sentiment, etc. Additionally, huge quantities of data are available in text corpora, suitable for their training in many domains and languages. This allows pretraining of LMs that can be adapted to many different target tasks.

ULMFiT uses the LSTM neural network with dropout to train language models on general text corpora. With the emphasis on transfer learning, ULMFiT consists of three phases, general-domain LM pretraining, target task LM fine-tuning, and target task classifier fine-tuning. The fine-tuning to the specifics of the target task starts with the trained LM. Training gradually reactivates different layers of the network to allow adaptation to the target domain. As different layers capture different types of information, they are fine-tuned to different extents, which is achieved by gradually decreasing the learning rates starting with the final network layer. The regular stochastic gradient descent (SGD) updates the model's parameters Θ at time step t, computed using the following formula:

$$\Theta_t = \Theta_{t-1} - \eta \bigtriangledown_\Theta J(\Theta),$$

where η is the learning rate, and $\bigtriangledown_\Theta J(\Theta)$ is the gradient of the model's objective function J with regard to parameters Θ. ULMFiT version of SGD adapts the learning rate in each layer of the neural network. For layer i, the parameters are set as follows:

$$\Theta_t^i = \Theta_{t-1}^i - \eta^i \bigtriangledown_{\Theta^i} J(\Theta),$$

where learning rate η^i is first tuned on the last layer of the network. Howard and Ruder (2018) recommend that learning rates of previous layers are geometrically decreased as $\eta^{i-1} = \frac{\eta^i}{2.6}$.

The learning rates are dynamically adjusted throughout the learning process with the idea that parameters shall quickly converge to a suitable region in the parameter space where they are fine-tuned. Rather than fine-tuning all the layers at once, which risks forgetting, ULMFit gradually unfreezes the model starting from the last layer and adds one more layer to fine-tuning in each epoch.

ULMFiT embeddings demonstrate good performance on many text classification tasks, including sentiment analysis, question classification and topic classification.

BERT (Bidirectional Encoder Representations from Transformers). BERT embeddings (Devlin et al. 2019) generalize the idea of language models (LMs) to masked language models, inspired by the gap-filling tests. The masked language model randomly masks some of the tokens from the input. The task of a LM is to predict each missing token based on its neighborhood. BERT uses the transformer architecture of neural networks (Vaswani et al. 2017) in a bidirectional sense. It introduces another task of predicting whether two sentences appear in a sequence. The input representation of BERT are sequences of tokens representing sub-word units. The input is constructed by summing the corresponding token, segment, and position embeddings.

Using BERT for classification requires adding connections between its last hidden layer and new neurons corresponding to the number of classes in the intended task. The fine-tuning process is typically applied to the whole network. All the BERT parameters and new class specific weights are fine-tuned jointly to maximize the log-probability of the correct labels.

BERT has shown excellent performance on 11 NLP tasks: 8 from the GLUE general language understanding evaluation benchmark suite (Wang et al. 2018), question answering, named entity recognition, and common-sense inference.

UniLM (Universal Language Model).   UniLM (Dong et al. 2019) extends the learning tasks of BERT (masked language model, and prediction if two sentences are consecutive) with the sequence to sequence prediction task. UniLM uses three learning tasks based on the masked input. Left-to-right unidirectional LM predicts the masked word from its left words (similarly to classical LMs). Right-to-left unidirectional LM predicts the masked word with the context of words on the right (equivalently to backward LMs). The bidirectional LM predicts the masked word based on the words on both the left and right.

The sequence-to-sequence task takes two texts and predicts a masked word in the second (target) sequence based on all the words in the first (source) sequence and the words left from it in the target sequence. For example, for the input sentence composed from tokens t_1 t_2 and the next sentence t_3 t_4 t_5, the input to the model is the following sequence:

$$[SOS]\ t_1\ t_2\ [EOS]\ t_3\ t_4\ t_5\ [EOS],$$

where [SOS] is the start of the sentence token, and [EOS] represents the end of the sentence token. Processing tokens t_1 and t_2, we have access to the first four tokens, including [SOS] and [EOS]. Token t_4 can access the first six tokens, but neither t_5 nor the last [EOS]. During training, tokens in both segments are randomly selected and replaced with the special [MASK] token. The model is trained to predict the masked tokens. The sequence to sequence model uses contiguous input text sequences during training, which encourages the model to learn the relationship between pairs of segments. To predict tokens in the second segment, UniLM learns to encode the first segment. In this way, the gap-filling task corresponds to the sequence-to-sequence LM and is composed of two parts, encoder and decoder. The encoder part is bidirectional, and the decoder is unidirectional (left-to-right). Similar to other encoder-decoder models, this can be exploited for text generation tasks such as summarization.

Like BERT, UniLM uses a deep transformer network with the self-attention mechanism and is meant to be fine-tuned with additional task-specific layers to adapt to various downstream tasks. Using different self-attention masks, UniLM aggregates different contexts and can be used for natural language understanding and generation tasks.

Experimental results of UniLM show that it improves over other text embeddings on several text understanding tasks, e.g., on the GLUE benchmark, in question answering tasks, in abstractive summarization of news and question generation.

The main difference between different approaches to contextual embeddings lies in the tasks they select for transfer learning, but they all use (generalized) language models. ULMFit (Howard and Ruder 2018) trains LSTM network on the problem of forward LM. ELMo (Peters et al. 2018) learns forward and backward unidirectional LM based on LSTM networks. A forward LM reads the text from left to right, and a backward LM encodes the text from right to left. BERT (Devlin et al. 2019) employs a bidirectional transformer encoder to fuse both the left and the right context to predict the masked words. BERT also explicitly models the relationship between a pair of texts, which has shown to be beneficial to some pairwise natural language understanding tasks, such as natural language inference. UniLM extends the masked LM of BERT with an additional sequence to sequence task applied to pairs of consecutive text segments. This task allows UniLM to encode the first segment and use it to predict (generate) the second segment.

The evolution of contextual text embedding shows that explicit extraction of numerical vectors is possible but not necessary. Instead of using the pretrained vectors, the whole network or a few layers of a pretrained network are used in learning a new task in the fashion of transfer learning. This has an advantage that layers of such a network can be fine-tuned to the target classification task. This is the predominant use of BERT and UniLM but is also used with ELMo and ULMFit.

3.4 Sentence and Document Embeddings

Like word embeddings (e.g., with methods like word2Vec, GloVE, ELMo, or BERT), embeddings can be constructed for longer units of texts such as sentences, paragraphs, and documents. Nevertheless, this is more challenging as the semantics of longer texts, which has to be encapsulated in the embedded vectors, is much more diverse and complex than the semantics of words.

Deep averaging network. The simplest approach is to average word embeddings of all words constituting a larger text unit such as a sentence or document. The deep averaging network by Iyyer et al. (2015) first averages the embeddings of words and bigrams. It uses them as an input to a feed-forward deep neural network to produce sentence embeddings.

doc2Vec. Le and Mikolov (2014) introduced the doc2Vec algorithm as an extension of word2vec, which usually outperforms simple averaging of word2vec vectors. The idea is to behave as if a document has another floating word-like vector, named doc-vector, contributing to all training predictions. The doc-vector is updated like other word-vectors and comes in two variants. Paragraph Vector–Distributed Memory (PV-DM) is analogous to the word2vec CBOW variant. The doc-vectors are obtained by training a neural network on the synthetic task of predicting a center word based on an average of both context word-vectors and the document's doc-vector. Paragraph Vector–Distributed Bag of Words (PV-DBOW) is analogous to word2Vec skip-gram variant. The doc-vectors are obtained by training a neural network on the synthetic task of predicting a target word just from the document's doc-vector. In training, this is combined with skip-gram testing, using both the doc-vector and nearby word-vectors to predict a single target word.

Transformer-based universal sentence encoder.   Similarly to the most successful word embedding models (e.g., BERT), recent successful sentence embedding models are based on the transformer architecture (Vaswani et al. 2017). Cer et al. (2018) have proposed the transformer-based universal sentence encoder that uses the encoding part of the transformer architecture. The encoder uses the attention mechanism to compute context-aware representations of words in a sentence that considers the ordering of tokens and the identity of other words. The learned word representations are converted to a fixed-length 512-dimensional sentence encoding vector by computing the element-wise sum of the representations at each word position.

The encoder of the universal sentence encoder by Cer et al. (2018) uses a transfer learning approach with three tasks.

Skip-thought task (Kiros et al. 2015)   can be used in the unsupervised learning from arbitrary running text (based on an input sentence). The task is to predict the previous and the next sentence around a given sentence.

Conversational input-response task   assures inclusion of parsed conversational data (Henderson et al. 2017), which uses pairs of email messages and their responses as the prediction task.

Classification task   for training on supervised data.

The sentence encoding demonstrated a strong performance on several NLP tasks (sentiment classification, subjectivity of sentences, question classification, semantic textual similarity) and has beaten other sentence representations on hate speech classification (Miok et al. 2019).

3.5   Cross-Lingual Embeddings

Most of the modern NLP approaches use word embeddings. The embedded vectors encode important information about word meaning, preserving semantic relations between words, which is even true across languages. Mikolov et al. (2013) noticed that dense word embedding spaces exhibit similar structures across languages, even when considering distant language pairs like English and Vietnamese. This means that embeddings independently produced from monolingual text resources can be aligned, resulting in a common cross-lingual representation, which allows for fast and effective integration of information in different languages. If we assume that the embeddings vectors are stored in a matrix, then we can use a transformation that approximately aligns vectors spaces of two languages L_1 and L_2, using:

$$W \cdot L_2 \approx L_1.$$

Especially for less-resourced languages, the mappings between languages can be of huge benefit as models trained on resourceful languages, such as English, can be transferred to less-resourced languages. The transferred models can be used in their original state (referred to as zero-shot transfer) or fine-tuned with a small amount of data in the target language (referred to as few-shot transfer). The idea of cross-lingual transfer of models is illustrated in Fig. 3.5.

The aligning transformations can categorized into three groups, listed below.

Supervised,   where the transformation is based on a set of aligned words, obtained from a dictionary.

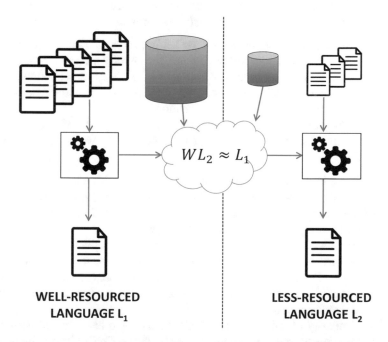

WELL-RESOURCED   **LESS-RESOURCED**
LANGUAGE L$_1$ **LANGUAGE L$_2$**

Fig. 3.5 Illustration of a cross-lingual transfer of a trained model from resource-rich language L_1 (on the left-hand side), where the model is trained on a large dataset, to less-resourced language L_2 (on the right-hand side), where the model is fine-tuned on a small dataset in the target language. Here W denotes the transformation matrix. Each language, L_1 and L_2, is first pretrained on a large text corpus (in blue) to form its monolingual representation, stored in the matrix L_1 and L_2, respectively

Semi-supervised,   where only a small seeding dictionary is initially provided. These initial aligned words serve to construct a mapping between languages, from which one can extract more matching words, and the procedure is iteratively repeated.

Unsupervised,   where the initial matching words are obtained either using the frequency of words in two languages, or a learning task, referred to as dictionary induction. The approach continues as in the semi-supervised case.

The success of cross-lingual embedding for model transfer is a convincing demonstration that transfer learning of representations really works, i.e. representations learned on one data set are transferred to other data sets. An in-depth discussion of cross-lingual embeddings methods, from words, sentences, and documents, to their evaluation and practical applications is presented by Søgaard et al. (2019).

3.6   Intrinsic Evaluation of Text Embeddings

As discussed in Sect. 1.4, the quality of data transformation approaches can be evaluated in two ways: extrinsic or intrinsic evaluation.

Extrinsic evaluation.   The most common setting for evaluating text embeddings is the extrinsic evaluation setting, where the quality of text embeddings is evaluated in some downstream learning task, such as classification. In this setting, the highest classification accuracy implicitly denotes the highest quality of the embeddings used in text transformation.

Intrinsic evaluation.   Intrinsic evaluation relates the produced embeddings to the similarity of entities in the original space. Below we list some common approaches.

- Similarity in the original space can be expressed with the *loss function* used in the data transformation task. For example, if a neural network is used to produce embeddings, the low values of its loss function indicate high similarity of input entities in the original space.
- In some cases, it is possible to reconstruct the original entities from the transformed ones, and the data *reconstruction error* can be used in the intrinsic evaluation. For example, autoencoder neural networks use the reconstruction error explicitly as the criterion to encode input entities in a latent lower-dimensional vector space.
- The similarity of entities in the original space is often assessed with a gold standard *human annotation*. For example, in the CoSimLex evaluation task (Armendariz et al. 2020), the similarity of words in different contexts is assessed by human annotators. The obtained scores are compared with the similarity of produced contextual word embeddings.

Intrinsic evaluation of embeddings was popularized by Mikolov et al. (2013), who designed a dedicated word analogy task for intrinsic evaluation of word embeddings.

Definition 3.1 (Word Analogy) The word analogy task is defined as follows. For a given relationship $a : b$, the task is to find a term y for a given term x so that the relationship between x and y best resembles the given relationship $a : b$ in the pair $x : y$.

There are two main groups of analogy categories: semantic and syntactic.

Semantic relationships. To illustrate a semantic relationship, consider e.g., the relationship 'capital of a country', i.e. 'city a is the capital of country b'. If the word pair $a : b$ is given as *Helsinki : Finland*, the task is to find the term y in the pair $x : y$, corresponding to the given relationship, e.g., *Stockholm : y, with the expected answer being y =* Sweden.

Syntactic relationships. In syntactic categories, each category refers to a grammatical feature, e.g., 'adjective a is the comparative form of adjective b'. The two words in any given pair have a common stem (or even the same lemma). For example, given the pair of adjectives in its comparative and the base form *longer : long*, the task is to find the term y corresponding to the relationship *darker : y*, with the expected answer being $y = dark$, i.e. the base form of the comparative form *darker*.

In the vector space, the analogy task is transformed into vector arithmetic. We search for nearest neighbors, i.e. we compute the distance between vectors, e.g., *d(vec(Helsinki), vec(Finland))*, and search for word y which would give the closest result in the expression *d(vec(Stockholm), vec(y))*. In the existing analogy datasets (Mikolov et al. 2013; Ulčar et al. 2020), the analogies are already prespecified, so one does not search for the closest result but checks if the prespecified word is indeed the closest; alternatively, one measures the distance between the given pairs. Many listed analogies will be matched if the relations from the dataset are correctly expressed in the embedding space. Therefore, as the evaluation metric, we can use the classification accuracy of the nearest neighbor classifier, where the query point is given as $vec(x) - vec(a) + vec(b)$.

The dataset proposed by Mikolov et al. (2013) is culturally biased towards the English language and the US culture (e.g., US football teams, US cities, or US states). Ulčar et al. (2020) redesigned this dataset to be culturally neutral and translated it into nine languages (Croatian, English, Estonian, Finnish, Latvian, Lithuanian, Russian, Slovenian, and Swedish). The resulting analogy tasks are composed of 15 categories: 5 semantic and 10 syntactic/morphological. No embedding method captures relations in all the categories, and the success is lower in morphologically rich languages.

To illustrate the range of tested relations, the categories are listed below (note that in some languages, certain categories do not make sense).

1. Capital cities in relation to countries, e.g., *Paris : France*.
2. A male family member in relation to an equivalent female member, e.g., *brother : sister*.
3. A non-capital city in relation to the country of this city, e.g., *Frankfurt : Germany*.

4. Species/subspecies in relation to their genus/familia, following not scientific (colloquial) terminology and relations, e.g., *salmon* : *fish*.
5. A city in relation to the river flowing through it, e.g., *London* : *Thames*.
6. An adverb in relation to the adjective it is formed from, e.g., *quiet* : *quietly*.
7. The morphologically derived opposite adjective in relation to the base form, e.g., *unjust* : *just*, or *dishonest* : *honest*.
8. The comparative form of adjective in relation to the base form, e.g., *longer* : *long*.
9. The superlative form of adjective in relation to the base form, e.g., *longest* : *long*.
10. The verbal noun (noun formed from a verb) in relation to the verb in the infinitive form, e.g., *sit* : *sitting*.
11. The nationality of the inhabitants of a country in relation to the country, e.g., *Albanians* : *Albania*.
12. The singular form of a noun in relation to the plural form of the noun, e.g., *computer* : *computers*; indefinite singular and definite plural are used in Swedish.
13. A genitive noun case in relation to the dative noun case in respective languages, e.g., in Slovene *ceste* : *cesti*.
14. A verb in 3rd person singular present tense form in relation to the verb in 3rd person singular past tense form, e.g., *goes* : *went*.
15. A verb in 3rd person singular in present tense form in relation to the verb in 3rd person singular in various tenses, e.g., *goes* : *gone*.

Cross-lingual word analogy task was proposed by Brychcín et al. (2019) as an intrinsic evaluation of cross-lingual embeddings. In this task, one pair of related words is in one language, and the other pair from the same category is in another language. For example, given the relationship in English *father* : *mother*, the task is to find the term y corresponding to the relationship *brat* (brother) : y in Slovene. The expected answer is $y = sestra$ (sister). Cross-lingual analogies are limited to the categories that participating languages have in common, i.e. certain syntactic categories are excluded. Ulčar et al. (2020) presented a dataset of consistent categories in all 72 combinations of the nine languages.

3.7   Implementation and Reuse

We selected the following methods to be demonstrated in reusable code snippets: matrix factorization embeddings LSA, topic embeddings LDA, neural embeddings word2vec, and contextual word embeddings BERT. Because of their popularity, several implementations exist. Our code is based on two popular programming libraries, *gensim* (Řehůřek and Sojka 2010) and *transformers* (Wolf et al. 2020), to ensure long-term usability.

3.7.1 LSA and LDA

The word embeddings method LSA (introduced in Sect. 3.2.5) and the topic embeddings method LDA (presented in Sect. 3.2.6) are demonstrated on the Reuters-21578 text categorization collection data set[2] (Apté et al. 1994), which contains a collection of documents that appeared on Reuters newswire in 1987. This popular document collection is suitable for our demonstration. The Jupyter notebook, which demonstrates the complete preprocessing phase and then applies LSA and LDA, is available in the repository of this monograph: https://github.com/vpodpecan/representation_learning/blob/master/Chapter3/LSA_LDA.ipynb.

3.7.2 word2vec

To demonstrate the word2vec word embedding method, described in Sect. 3.3.1. we use the pre-trained English vectors built on the English CoNLL17 corpus available in the NLPL word embeddings repository[3] (Fares et al. 2017). The famous analogy vec(king) − vec(man) + vec(woman) ≈ vec(queen) is demonstrated using helper functions from the *gensim* library and a toy corpus is used to demonstrate how to compute document similarities using word2vec. The Jupyter notebook is available in the repository of this monograph: https://github.com/vpodpecan/representation_learning/blob/master/Chapter3/word2vec.ipynb.

3.7.3 BERT

The pretrained BERT model, presented in Sect. 3.3.3, is demonstrated on the IMDB movie review dataset (Maas et al. 2011) with sentiment analysis. We compare three approaches:

- TF-IDF with SVM classifier is used as a baseline;
- the pre-trained DistilBERT from the *transformers* library uses fewer resources than BERT and is already prepared for sentiment classification;
- the BERT model directly fine-tuned on the IMDB reviews.

The Jupyter notebook is available in the repository of this monograph: https://github.com/vpodpecan/representation_learning/blob/master/Chapter3/BERT.ipynb.

---

[2]https://archive.ics.uci.edu/ml/datasets/Reuters-21578+Text+Categorization+Collection.
[3]http://vectors.nlpl.eu/repository/.

References

Chidanand Apté, Fred Damerau, and Sholom M. Weiss. Automated learning of decision rules for text categorization. *ACM Transactions on Information Systems*, 12(3):233–251, 1994.

Carlos S. Armendariz, Matthew Purver, Matej Ulčar, Senja Pollak, Nikola Ljubešić, Marko Robnik-Šikonja, Mark Granroth-Wilding, and Kristiina Vaik. CoSimLex: A resource for evaluating graded word similarity in context. In *Proceedings of the 12th Language Resources and Evaluation Conference (LREC)*, pages 5880–5888, 2020.

David M. Blei, Andrew Y. Ng, and Michael I. Jordan. Latent Dirichlet Allocation. *Journal of Machine Learning Research*, 3:993–1022, 2003.

Tomáš Brychcín, Stephen Taylor, and Lukáš Svoboda. Cross-lingual word analogies using linear transformations between semantic spaces. *Expert Systems with Applications*, 2019.

John A. Bullinaria and Joseph P. Levy. Extracting semantic representations from word co-occurrence statistics: A computational study. *Behavior research methods*, 39(3):510–526, 2007.

Rich Caruana. Multitask learning. *Machine Learning*, 28:41–75, 1997.

Daniel Cer, Yinfei Yang, Sheng-yi Kong, Nan Hua, Nicole Limtiaco, Rhomni St. John, Noah Constant, Mario Guajardo-Cespedes, Steve Yuan, Chris Tar, Brian Strope, and Ray Kurzweil. Universal sentence encoder for English. In *Proceedings of the 2018 Conference on Empirical Methods in Natural Language Processing: System Demonstrations*, pages 169–174, 2018.

Franca Debole and Fabrizio Sebastiani. Supervised term weighting for automated text categorization. In *Text Mining and its Applications*, pages 81–97. Springer, 2004.

Scott Deerwester, Susan T. Dumais, George W. Furnas, Thomas K. Landauer, and Richard Harshman. Indexing by latent semantic analysis. *Journal of the American Society for Information Science*, 41(6):391–407, 1990.

Jacob Devlin, Ming-Wei Chang, Kenton Lee, and Kristina Toutanova. BERT: Pre-training of deep bidirectional transformers for language understanding. In *Proceedings of the 2019 Conference of the North American Chapter of the Association for Computational Linguistics: Human Language Technologies, Volume 1 (Long and Short Papers)*, pages 4171–4186, 2019.

Li Dong, Nan Yang, Wenhui Wang, Furu Wei, Xiaodong Liu, Yu Wang, Jianfeng Gao, Ming Zhou, and Hsiao-Wuen Hon. Unified language model pre-training for natural language understanding and generation. In *Advances in Neural Information Processing Systems*, pages 13063–13075, 2019.

Murhaf Fares, Andrey Kutuzov, Stephan Oepen, and Erik Velldal. Word vectors, reuse, and replicability: Towards a community repository of large-text resources. In *Proceedings of the 21st Nordic Conference on Computational Linguistics*, pages 271–276, 2017.

Ronen Feldman and James Sanger. *The Text Mining Handbook: Advanced Approaches in Analyzing Unstructured Data*. Cambridge University Press, 2006.

John Rupert Firth. *Papers in Linguistics, 1934–1951*. Oxford University Press, 1957.

Zellig S. Harris. Distributional structure. *Word*, 10(2–3):146–162, 1954.

Matthew L. Henderson, Rami Al-Rfou, Brian Strope, Yun-Hsuan Sung, László Lukács, Ruiqi Guo, Sanjiv Kumar, Balint Miklos, and Ray Kurzweil. Efficient natural language response suggestion for smart reply. *arXiv*, abs/1705.00652, 2017.

Thomas Hofmann. Probabilistic Latent Semantic Indexing. In *Proceedings of the 22nd Annual International ACM SIGIR Conference on Research and Development in Information Retrieval*, pages 50–57, 1999.

Jeremy Howard and Sebastian Ruder. Universal language model fine-tuning for text classification. In *Proceedings of the 56th Annual Meeting of the Association for Computational Linguistics (Volume 1: Long Papers)*, pages 328–339, 2018.

Mohit Iyyer, Varun Manjunatha, Jordan Boyd-Graber, and Hal Daumé III. Deep unordered composition rivals syntactic methods for text classification. In *Proceedings of the 53rd Annual Meeting of the Association for Computational Linguistics and the 7th International Joint Conference on Natural Language Processing (Volume 1: Long Papers)*, pages 1681–1691, 2015.

Karen Spärck Jones. A statistical interpretation of term specificity and its application in retrieval. *Journal of Documentation*, 28:11–21, 1972.

Ryan Kiros, Yukun Zhu, Russ R. Salakhutdinov, Richard Zemel, Raquel Urtasun, Antonio Torralba, and Sanja Fidler. Skip-thought vectors. In *Advances in Neural Information Processing Systems*, pages 3294–3302, 2015.

Man Lan, Chew Lim Tan, Jian Su, and Yue Lu. Supervised and traditional term weighting methods for automatic text categorization. *IEEE Transactions on Pattern Analysis and Machine Intelligence*, 31(4):721–735, 2009.

Quoc Le and Tomas Mikolov. Distributed representations of sentences and documents. In *Proceedings of International Conference on Machine Learning*, pages 1188–1196, 2014.

Andrew L. Maas, Raymond E. Daly, Peter T. Pham, Dan Huang, Andrew Y. Ng, and Christopher Potts. Learning word vectors for sentiment analysis. In *Proceedings of the 49th Annual Meeting of the Association for Computational Linguistics: Human Language Technologies*, pages 142–150, 2011.

Christopher D. Manning and Hinrich Schütze. *Foundations of Statistical Natural Language Processing*. The MIT Press, Cambridge, Massachusetts, 1999.

Justin Martineau and Tim Finin. Delta TFIDF: An improved feature space for sentiment analysis. In *Proceedings of the Third AAAI International Conference on Weblogs and Social Media*, 2009.

Tomas Mikolov, Ilya Sutskever, Kai Chen, Greg S. Corrado, and Jeff Dean. Distributed representations of words and phrases and their compositionality. In *Advances in neural information processing systems*, pages 3111–3119, 2013.

Kristian Miok, Dong Nguyen-Doan, Blaž Škrlj, Daniela Zaharie, and Marko Robnik-Šikonja. Prediction uncertainty estimation for hate speech classification. In *Proceedings of the International Conference on Statistical Language and Speech Processing*, pages 286–298, 2019.

Jeffrey Pennington, Richard Socher, and Christopher Manning. GloVe: Global vectors for word representation. In *Proceedings of Empirical Methods in Natural Language Processing, EMNLP*, pages 1532–1543, 2014.

Matthew Peters, Mark Neumann, Mohit Iyyer, Matt Gardner, Christopher Clark, Kenton Lee, and Luke Zettlemoyer. Deep contextualized word representations. In *Proceedings of the 2018 Conference of the North American Chapter of the Association for Computational Linguistics: Human Language Technologies, Volume 1 (Long Papers)*, pages 2227–2237, 2018.

Lorien Y Pratt, Jack Mostow, Candace A Kamm, and Ace A Kamm. Direct transfer of learned information among neural networks. In *Proceedings of AAAI*, pages 584–589, 1991.

Radim Řehůřek and Petr Sojka. Software framework for topic modelling with large corpora. In *Proceedings of the LREC 2010 Workshop on New Challenges for NLP Frameworks*, pages 45–50, 2010.

Stephen E. Robertson and Steve Walker. Some simple effective approximations to the 2-Poisson model for probabilistic weighted retrieval. In *Proceedings of the 17th Annual International ACM SIGIR Conference on Research and Development in Information Retrieval*, pages 232–241, 1994.

Gerard Salton and Christopher Buckley. Term-weighting approaches in automatic text retrieval. *Information Processing and Management*, 24(5):513–523, 1988.

Anders Søgaard, Ivan Vulić, Sebastian Ruder, and Manaal Faruqui. *Cross-Lingual Word Embeddings*, volume 12-2. Morgan & Claypool Publishers, 2019.

Matej Ulčar, Kristiina Vaik, Jessica Lindström, Milda Dailidėnaitė, and Marko Robnik-Šikonja. Multilingual culture-independent word analogy datasets. In *Proceedings of the 12th Language Resources and Evaluation Conference*, pages 4067–4073, 2020.

Ashish Vaswani, Noam Shazeer, Niki Parmar, Jakob Uszkoreit, Llion Jones, Aidan N Gomez, Łukasz Kaiser, and Illia Polosukhin. Attention is all you need. In *Advances in Neural Information Processing Systems*, pages 5998–6008, 2017.

Alex Wang, Amanpreet Singh, Julian Michael, Felix Hill, Omer Levy, and Samuel Bowman. GLUE: A multi-task benchmark and analysis platform for natural language understanding. In *Proceedings of the 2018 EMNLP Workshop BlackboxNLP: Analyzing and Interpreting Neural Networks for NLP*, pages 353–355, 2018.

Thomas Wolf, Lysandre Debut, Victor Sanh, Julien Chaumond, Clement Delangue, Anthony Moi, Pierric Cistac, Tim Rault, Rémi Louf, Morgan Funtowicz, Joe Davison, Sam Shleifer, Patrick von Platen, Clara Ma, Yacine Jernite, Julien Plu, Canwen Xu, Teven Le Scao, Sylvain Gugger, Mariama Drame, Quentin Lhoest, and Alexander M. Rush. Transformers: State-of-the-art natural language processing. In *Proceedings of the 2020 Conference on Empirical Methods in Natural Language Processing: System Demonstrations*, pages 38–45. Association for Computational Linguistics, 2020.

Chapter 4
Propositionalization of Relational Data

Relational learning addresses the task of learning models or patterns from relational data. Complementary to relational learning approaches that learn directly from relational data, developed in the Inductive Logic Programming research community, this chapter addresses the propositionalization approach of first transforming a relational database into a single-table representation, followed by a model or pattern construction step using a standard machine learning algorithm. This chapter outlines several propositionalization algorithms for transforming relational data into a tabular data format. The chapter is structured as follows. Section 4.1 presents the problem of relational learning, including its definition and the main approaches. Relational data representations and an illustrative toy example, which we expand in further sections, are introduced in Sect. 4.2. Section 4.3 presents propositionalization and briefly outlines feature construction approaches used in standard propositionalization algorithms. Section 4.4 outlines selected propositionalization approaches. Section 4.5 describes an efficient and practically applicable propositionalization method, named wordification. Section 4.6 presents an approach to training deep neural networks on propositionalized relational data, named Deep Relational Machines (DRM). The chapter concludes by presenting selected propositionalization methods implemented in Jupyter Python notebooks in Sect. 4.7.

4.1 Relational Learning

Standard machine learning and data mining algorithms induce models or patterns from a given data table, where each example corresponds to a single row, i.e. a single fixed-length attribute-value tuple. These learners, which can be referred to as *propositional learners*, thus use as input a propositional representation. However, the restriction to a single table poses a challenge for data that are naturally represented in a relational representation, referred to as *relational data* or *multi-*

© Springer Nature Switzerland AG 2021   83
N. Lavrač et al., *Representation Learning*,
https://doi.org/10.1007/978-3-030-68817-2_4

relational data representation. *Relational learning* problems cannot be directly represented with a tabular representation without a loss of information. These problems, which include the analysis of complex structured data, are naturally represented using multiple relations.

Learning from relational data can be approached in two main ways:

Relational learning and inductive logic programming. These approaches, known as Relational Learning (RL) (Quinlan 1990), Inductive Logic Programming (ILP) (Muggleton 1992; Lavrač and Džeroski 1994; Srinivasan 2007), Relational Data Mining (RDM) (Džeroski and Lavrač 2001), Statistical Relational Learning (SRL) (Getoor 2007), learn a relational model or a set of relational patterns *directly from the relational data.*

Propositionalization. Propositionalization techniques transform a relational representation into a propositional single-table representation by constructing complex features (Lavrač et al. 1991; Kramer et al. 2000; Krogel et al. 2003; Železný and Lavrač 2006; Kuželka and Železný 2011), and then use a propositional learner on the transformed data table.

The former strategy requires the design of dedicated algorithms for analyzing relational data, while the latter strategy involves a data preprocessing step, named *propositionalization*, enabling the user to transform the relational data into a tabular data format and subsequently employ a range of standard propositional learning algorithms on the transformed tabular data representation. Propositionalization, which is one of the two central topics of this monograph, is discussed in this chapter.

4.2 Relational Data Representation

Depending on the context, one can use different notations for logical rules:

Explicit. This notation explicitly marks the condition (IF) part of the rule, the conjunction (AND) of features in the condition part of the rule, and the conclusion (THEN) part. An example rule described in this formalism is shown in the first line of Table 4.1.

Formal. This notation is based on formal (propositional) logic. The implication sign ($\leftarrow$) is typically written from right to left (as in Prolog, see below), but it may also appear in the other direction. It will be used when the theoretical

Table 4.1 An example rule described in three rule representation formalisms, where $\wedge$ denotes the conjunction operator, $\leftarrow$ and : - denote the logical implication, and X denotes a logical variable (in Prolog, names of variables start with capital letters, while constants start with lowercase letters)

Notation	Example rule
Explicit	IF *Shape* = *triangle* $\wedge$ *Color* = *red* $\wedge$ *Size* = *big* THEN *Class* = *positive*
Formal	*Class* = *positive* $\leftarrow$ *Shape* = *triangle* $\wedge$ *Color* = *red* $\wedge$ *Size* = *big*
Logical	`positive(X):-shape(X,triangle),color(X,red),size(X,big).`

properties of rule learning algorithms are discussed. The example rule described
in this formalism is shown in the second line of Table 4.1.

Logical (Prolog). The syntax of the logic programming language Prolog (Bratko
1990), which is most frequently used in ILP , will mostly be used in the relational
learning examples in this chapter. The example rule described in this formalism
is shown in the third line of Table 4.1.

This section presents an illustrative example shown in Sect. 4.2.1, followed by
the presentation of two data description formats of particular interest: logical representation in the Prolog format in Sect. 4.2.2, and relational database representation
in Sect. 4.2.3.

4.2.1 Illustrative Example

The illustrative example used in this chapter is Michalski's East-West trains
challenge dataset (Michie et al. 1994), illustrated in Fig. 4.1, where the goal of
the learned model is to classify the direction of an unseen train. The data consists
of ten examples, i.e. ten trains t1, ..., t10, where the predicates eastbound and
westbound indicate the class, i.e. whether the train is eastbound or westbound.

eastbound: eastbound(t1).eastbound(t2).... eastbound(t5).
westbound: westbound(t6).westbound(t7).... westbound(t10).

4.2.2 Example Using a Logical Representation

In a logic programming representation of the problem using the Prolog programming language, each train is described by a set of ground facts, each describing the

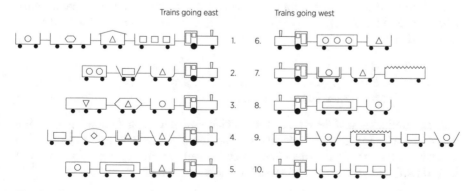

Fig. 4.1 The ten-train East-West trains challenge

Table 4.2 Prolog representation of the first train in the East-West trains dataset. The predicate cnumber contains the number of cars for a given train, hasCar connects a car to a given train, cshape, clength, croof, and cwheels define properties of a given car, hasLoad associates a given load with the given car, and lshape and lnumber describe properties of a given load

```
eastbound(t1).

cnumber(t1,4).

hasCar(t1,c11).                hasCar(t1,c12).
cshape(c11,rectangular).       cshape(c12,rectangular).
clength(c11,long).             clength(c12,short).
croof(c11,no_roof).            croof(c12,peak).
cwheels(c11,2).                cwheels(c12,2).
hasLoad(c11,l11).              hasLoad(c12,l12).
lshape(l11,rectangular).       lshape(l12,triangular).
lnumber(l11,3).                lnumber(l12,1).

hasCar(t1,c13).                hasCar(t1,c14).
cshape(c13,rectangualar).      cshape(c14,rectangular).
clength(c13,long).             clength(c14,short).
croof(c13,no_roof).            croof(c14,no_roof).
cwheels(c13,3).                cwheels(c14,2).
hasLoad(c13,l13).              hasLoad(c14,l14).
lshape(l13,hexagonal).         lshape(l14,circular).
lnumber(l13,1).                lnumber(l14,1).
```

properties of a train and its cars. The set of Prolog facts describing the first train is given in Table 4.2.

In this example, each train consists of 2–4 cars. The cars have properties like shape (rectangular, oval, u-shaped), length (long, short), number of wheels (2, 3), type of roof (no_roof, peaked, jagged), shape of load (circle, triangle, rectangle), and number of loads (1–3).

For this small dataset consisting of ten instances, a relational learner can induce the following Prolog rule distinguishing between eastbound and westbound trains, which states that 'A train T is eastbound if it contains a short car with a roof', expressed as follows:

```
eastbound(T) :-
    hasCar(T,C), clength(C,short),
    not croof(C,no_roof).
```

This rule indicates, that the learned Prolog clauses can use negation in the condition part of the rule. In the rest of this chapter, for simplicity, we will use the expression croof(C,closed), i.e. that car C has a 'closed car' for a car with a peaked or jagged roof, expressed in the rule condition as not croof(C,no_roof).

4.2.3   Example Using a Relational Database Representation

In a relational database, a given problem is presented as a set of relations $\{R_1, \ldots, R_n\}$ and a set of foreign-key connections between the relations denoted by $R_i \rightarrow R_j$, where R_i has a foreign-key pointing to relation R_j. The foreign-key connections correspond to the relationships in an Entity-Relationship (ER) diagram.

Take the ER diagram, illustrated in Fig. 4.2, which represents the data model describing the structure of the East-West trains challenge data. Instead of the data representation in the form of Prolog facts shown in Table 4.2, the data can be stored in a relational database, consisting of three relational tables presented in Table 4.3.

Entity-relationship diagrams can be used to choose a proper logical representation for the data. If we store the data in a relational database, the most obvious representation is to have a separate table for each entity in the domain, with relationships being expressed by foreign keys. This is not the only possibility: for instance since the relationship between Car and Load is one-to-one, this relationship serves just as 'syntactic sugar' as both entities could be combined in a single table, while entities linked by a one-to-many relationship cannot be combined without either introducing significant redundancy or a significant loss of information, e.g., introduced through aggregate attributes. Note that one-to-many relationships distinguish relational learning and Inductive Logic Programming from propositional learning.

The ER diagram of Fig. 4.2 shows three relations appearing in the East-West train challenge: the Train, Car and the Load relational tables. The boxes in the ER diagram indicate *entities*, which are individuals or parts of individuals: the Train entity is the individual, each Car is part of a train, and each Load is part of a car. The ovals denote attributes of entities. The diamonds indicate *relationships* between

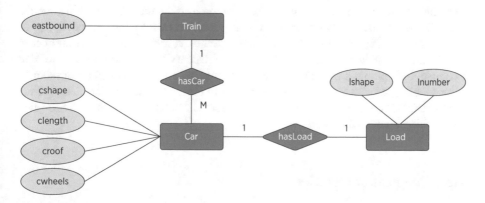

Fig. 4.2 Entity-relationship diagram for the East-West trains challenge, including a *one-to-many relationship* hasCar from Train to Car (corresponding to the hasCar(T,C) predicate), and a *one-to-one relationship* hasLoad between Car and Load (corresponding to the hasLoad(C,L) predicate)

Table 4.3 A relational database representation of the East-West trains challenge dataset

TRAIN	
trainID	Eastbound
t1	True
:	:
t5	True
t6	False
:	:

LOAD			
loadID	lshape	lnumber	carID
111	Rectangular	3	c14
112	Triangular	1	c13
113	Hexagonal	1	c12
114	Circular	1	c11
:	:	:	:

CAR					
carID	cshape	clength	croof	cwheels	trainID
c11	Rectangular	Long	no_roof	2	t1
c12	Rectangular	Short	Peak	2	t1
c13	Rectangular	Long	no_roof	3	t1
c14	Rectangular	Short	no_roof	2	t1
:	:	:	:	:	:

entities. There is a *one-to-many relationship* from Train to Car, indicating that each train can have an arbitrary number of cars, but each car is contained in exactly one train; and a *one-to-one relationship* between Car and Load, indicating that each car has exactly one load and each load is a part of exactly one car.

In the East-West trains challenge data, instead of the data representation in the form of Prolog facts shown in Table 4.2, the data can be stored in a relational database, consisting of three relational tables presented in Table 4.3. In a database terminology, a *primary key* of a table is a unique identifier of each record. For example, trainID is the primary key for records in the TRAIN table, while carID and loadID are primary keys in the CAR and LOAD tables, respectively. A *foreign key* is a key used to link two tables together. A foreign key is a column or a combination of columns whose values match a primary key in a different table. For example, the column trainID in the CAR table is its foreign key, linking records in this table (i.e. cars) to their train, described in the TRAIN table. The same is true for the carID column in the LOAD table, where carID plays a role of the foreign key, linking loads to their cars.

4.3   Propositionalization

Relational learners that directly use the relational data representations intertwine feature construction and model construction. On the other hand, in proposition-alization, these two steps are separated. The workload of finding good relational features is performed by the propositionalization algorithm, while the work of combining these features to produce a good model is offloaded to the propositional learner having its own hypothesis language bias, e.g., decision trees, classification rules, Support Vector Machines (SVM), or more recently, deep neural networks. Importantly, propositionalization can be performed with many Machine Learning (ML) or Data Mining (DM) tasks in mind: classification, regression, association discovery, clustering, etc. Nevertheless, in this chapter, we decided to focus on the classification task, which is the most frequently addressed task.

In propositionalization, we use entity-relationship diagrams to define types of objects in the domain, where each entity corresponds to a distinct type. The data model constitutes a *language bias* that can be used to restrict the hypothesis space (i.e. the space of possible models) and guide the search for good models. In most problems, only individuals and their parts exist as entities, which means that the entity-relationship model has a tree-structure with the individual entity at the root and only *one-to-one* or *one-to-many* relations in the downward direction (i.e. not containing any *many-to-many* relations). Representations with this restriction are referred to as *individual-centered representations* (Flach and Lachiche 1999).

Transformation of relational data into a single table format through proposition-alization is a suitable approach to relational learning only when the problem at hand is *individual-centered*, i.e. when learning occurs only at the level of the individual. An example dataset that complies with the individual-centered representation bias is the East-West trains dataset (recall its ER diagram in Fig. 4.2 and parts of the data in Tables 4.2 and 4.3). Take as another example a problem of classifying authors into research fields given a citation network: the author is the individual and learning is performed at the author level (e.g., assigning class labels to authors).

If some loss of information is acceptable for the user, propositionalization can be achieved, e.g., by using aggregation queries on the relational database to compress the data into a single table format. Feature construction through aggregation queries has been addressed in the early work on propositionalization of Kramer et al. (2001) and Krogel et al. (2003), which includes an extensive overview of different feature construction approaches. Alternative propositionalization approaches per-form relational feature construction, which is the most common approach to data transformation, including relational subgroup discovery system RSD (Železný and Lavrač 2006) briefly outlined in Sects. 4.3.2 and 4.3.3 below.

4.3.1 Relational Features

Most of the propositionalization algorithms, including the relational subgroup discovery algorithm RSD (Železný and Lavrač 2006), use ILP notation for features and rules. Individual features are described by *literals* or *conjunctions of literals*. For example, in the toy example of Sect. 4.2, the features describing the cars are either literals (e.g., `clength(C,short)`), or conjunctions of literals (e.g., `clength(C,short), not croof(C,no_roof)`). For simplicity, the latter conjunction will be expressed as `clength(C,short), croof(C,closed)` in the rest of this section.

Given the training examples and the background knowledge, an ILP learner can induce the following two Prolog rules describing eastbound trains.

```
eastbound(T) :-
    hasCar(T,C), clength(C,short),
    croof(C,closed).

eastbound(T) :-
    hasCar(T,C1), clength(C1,short),
    hasCar(T,C2), croof(C2,closed).
```

In the rules, T is a global, universally quantified variable denoting any train, while C, C1 and C2 are existentially quantified local variables introduced in the rule body, denoting the existence of a car with a certain property.

- The first rule (which is the same as in the example of Sect. 4.2.2) expresses that for every train T, IF there exists (at least one) short car C, which is closed (has a roof), THEN train T is eastbound. This means that all trains that have a short closed car are going East.
- The second rule states that trains with (at least one) short car and (at least one, possibly different) closed car are going East. The second rule is more general than the first, covering all the instances covered by the first rule. Besides, it covers the instances where the closed car is different from the short car.

In the terminology used in this chapter, we say that the body of the first rule consists of *one relational feature*, while the body of the second rule contains *two relational features*. The main characteristic of relational features is that they localize variable sharing: the only variable shared among features is the global variable occurring in the rule head. Relational features are formally defined as follows.

Definition 4.1 (Relational Feature) A relational feature is a minimal set of literals such that no local (existential) variable occurs both inside and outside the set of literals.

To illustrate relational feature construction, let us take the rule containing two distinct relational features.

```
eastbound(T) :-
    hasCar(T,C1), clength(C1,short),
    hasCar(T,C2), not croof(C2,no_roof).
```

Manual construction of relational features could be performed by identifying and naming the relational features by two separate predicate names `hasShortCar` and `hasClosedCar`. The two predicates could be defined as follows.

```
hasShortCar(T) :- hasCar(T,C), clength(C,short).
```

```
hasClosedCar(T)  :- hasCar(T,C), not croof(C,no_roof).
```

Using these two manually constructed features as background predicates, the second rule can be translated into a rule without local variables:

```
eastbound(T)  :- hasShortCar(T), hasClosedCar(T).
```

This rule states that IF a train has a short car AND has a closed car THEN the train is eastbound. Notably, this rule refers only to the properties of trains, and hence can be expressed extensionally by a single table describing trains in terms of these properties. In the following, we will see how we can automatically construct such relational features and the corresponding propositional table.

4.3.2   Automated Construction of Relational Features by RSD

The actual goal of propositionalization is to automatically generate several relevant relational features about an individual that the learner can use to construct a model. In the machine learning literature, *automated feature construction* is also known as *constructive induction*. Propositionalization is thus a form of constructive induction since it involves changing the representation for learning.

As manual feature construction is complex and/or unfeasible, the task of propositionalization is to automate the construction of features, acting as queries about each individual. These queries will be evaluated as *true* or *false* on the original data when constructing a transformed data table. They form truth values of constructed relational features (see Sect. 4.3.3).

Let us briefly describe an approach to automated relational feature construction implemented in the relational subgroup discovery system RSD (Železný and Lavrač 2006), representing one of the most common approaches to propositionalization.

Formally, in accordance with Definition 4.1, a *relational feature*, which expresses an individual's property by a conjunction of predicates and properties, is composed as follows.

- There is exactly one free variable that plays the role of the global variable in rules. This variable corresponds to the individual's identifier in the individual-centered representation (e.g., for the trains example, the global variable T represents any individual train).
- Each predicate introduces a new (existentially quantified) local variable, and uses either the global variable or one of the local variables introduced by a predicate in one of the preceding literals, e.g., the predicate hasCar(T,C) introduces a new local existential variable C for a car of train T.
- Properties do not introduce new variables, e.g., lshape(L,triangular) stands for a triangular shape of load L, provided that load variable L has been already instantiated.
- All variables are 'consumed', i.e. used either by a predicate or a property.

Relational feature construction can be restricted by parameters that define the maximum number of literals constituting a feature, number of local variables, and number of occurrences of individual predicates. For example, a relational feature denoting the property of a train having a car with a triangular load, can be constructed for the trains example, in the language bias allowing for up to 4 literals and 3 variables:

```
f200(T) :-
    hasCar(T,C), hasLoad(C,L), lshape(L,triangular).
```

This example shows that a typical feature contains a chain of predicates, closed off by one or more properties.

Properties can also establish relations between parts of the individual, e.g., a feature that expresses the property of 'having a car whose shape is the same as the shape of its load':

```
f300(T) :-
    hasCar(T,C), cshape(C,CShape),
    hasLoad(C,L), lshape(L,LShape),
    CShape = LShape.
```

Note that the language bias in which these illustrative features were generated, allowing for up to 4 literals and 3 variables, is richer than the language bias in Sect. 4.3.3, where we constrain the construction of features to those that use at most two predicates that introduce local variables, and at most two predicates that describe properties.

For illustration, within a default RSD language bias, defined by constraining feature construction with the default parameters (clauselength=8, negation=none, min_coverage=1, filtering=true, depth=4), RSD generates 116 features, from the simplest feature:

```
f1(T) :- hasCar(T,C), carPosition(C,4).
```

to the most complex feature:

```
f116(T) :-
    hasCar(T,C), carRoof(C,flat),
    carLoadShape(C,circle), carLoadNum(C,1)}.
```

4.3.3 Automated Data Transformation and Learning

Once a set of relational features has been automatically constructed, these features start acting as database queries q_i that return *true* (value 1) or *false* (value 0) for a given individual. In the illustrative example of classifying trains as eastbound or westbound, the generated features are evaluated for each individual train. In this way, a transformed data table is constructed.

Table 4.4 The propositionalized tabular form of the East-West trains dataset

T	f1(T)	f2(T)	f3(T)	f4(T)	...	f116(T)	eastbound(T)
t1	1	0	0	1	...	0	true
t2	1	1	0	1	...	1	true
:	:	:	:	:	:	:	:
t6	0	0	1	1	...	1	false
t7	1	0	0	0	...	0	false
:	:	:	:	:	:	:	:

Transformed data table.   We demonstrate data transformation on the East-West trains dataset, using relational feature construction algorithm RSD (Železný and Lavrač 2006). Each constructed feature describes a property of a train, where a feature can involve a rather complex query involving multiple relations, as long as the query returns either value 1 (i.e. *true*) or 0 (i.e. *false*) when evaluated on a given train. The target attribute value (i.e. the train direction) is not preprocessed (instead of being evaluated, it is just appended to the transformed feature vectors). A simplified output of RSD transformation is shown in Table 4.4.

Learning from transformed data table.   In model construction from the propositionalized data table, the learner exploits the feature values to construct a classification model. If a decision tree learner is used, each node in the tree then contains a feature and has two branches (*true* and *false*). To classify unseen individuals, the classifier evaluates the features found in the decision tree nodes. It follows the branches according to their answers to arrive at a classification in a tree leaf. Alternatively, learning can be performed by a classification rule learning algorithm. E.g., taking Table 4.4 as input (using column eastbound(T) as the class attribute), an attribute-value rule learner such as CN2 (Clark and Niblett 1989) could be used to induce if-then rules such as

eastbound(T)=true ← f1(T)=1 ∧ f4(T)=1,

where f1(T) and f4(T) are two of the automatically generated features, and ∧ is the conjunction operator. On the other hand, if the target attribute were disregarded, a clustering or association rule learning algorithm could discover interesting patterns in the data.

4.4   Selected Propositionalization Approaches

In propositionalization, relational feature construction followed by feature evaluation and table construction is the most common approach to data transformation. LINUS (Lavrač et al. 1991) was one of the pioneering propositionalization approaches using automated relational feature construction. LINUS was restricted to the generation of features that do not allow recursion and existential local variables, which means that the target relation could not be many-to-many and self-referencing. The second limitation was more serious: the queries could not contain joins (conjunctions of literals). The LINUS approach had many followers, including SINUS (Lavrač and Flach 2001), and relational subgroup discovery system RSD (Železný and Lavrač 2006).

Aggregation approaches to propositional feature construction are a popular alternative to relational feature construction. An *aggregate* can be viewed as a function that maps a set of records in a relational database to a single value, which can be done by adding a constructed attribute to one or to several tables of the relational database. If an aggregation function considers values from only one table of a relational database, the aggregation is equivalent to feature construction. For instance, in the database representation of trains example in Table 4.3, one could add a binary aggregate column to the CAR table by testing the predicate '#cwheels is odd'. Aggregation becomes more powerful if it involves more tables in the relational database. An example of a first-order aggregate would be a column added to the TRAIN table, which computes the total number of wheels on all cars of the given train; this requires information from the CAR table. A second-order aggregate (encompassing all tables) could add a column to the TRAIN table representing its total number of wheels on cars containing rectangular-shaped loads.

A selection of propositionalization approaches is outlined below. An interested reader can find extensive overviews of different feature construction approaches in the work of Kramer et al. (2001) and Krogel et al. (2003).

Relaggs (Krogel and Wrobel 2001)   stands for relational aggregation. It is a propositionalization approach that takes the input relational database schema as a basis for declarative bias, using optimization techniques usually used in relational databases (e.g., indexes). The approach employs aggregation functions to summarize non-target relations concerning the individuals in the target table. Safarii (Knobbe 2005) is a commercial multi-relational data mining tool, which offers a unique pattern language that combines ILP-style descriptions with aggregates. Its tool ProSafarii offers several preprocessing utilities, including propositionalization via aggregation.

Stochastic propositionalization (Kramer et al. 2000)   employs a search strategy similar to random mutation hill-climbing: the algorithm iterates over generations of individuals, which are added and removed with a probability proportional to the fitness of individuals, where the fitness function used is based on the Minimum Description Length (MDL) principle (Rissanen 1983).

1BC (Flach and Lachiche 1999)   uses the propositional naive Bayes classifier to handle relational data. It first generates a set of first-order conditions and then uses them as attributes in the naive Bayes classifier. However, the transformation is done dynamically, as opposed to standard propositionalization, which is performed as a static step of data preprocessing. This approach is extended by 1BC2 (Lachiche and Flach 2003), which allows distributions over sets, tuples, and multisets, thus enabling the naive Bayes classifier to consider also structured individuals.

Tertius (Flach and Lachiche 2001)   is a top-down rule discovery system, incorporating first-order clausal logic. The main idea is that no particular prediction target is specified beforehand. Hence, Tertius can be seen as an ILP system that learns rules in an unsupervised manner. Its relevance for this survey lies in the fact that Tertius encompasses 1BC, i.e. relational data is handled through 1BC transformation.

RSD (Železný and Lavrač 2006)   is a relational subgroup discovery algorithm composed of two main steps: the propositionalization step and the (optional) subgroup discovery step. The output of the propositionalization step can also be used as input to other propositional learners. Using different biases, RSD efficiently produces an exhaustive list of first-order features that comply with the user-defined mode constraints, similar to those of Progol (Muggleton 1995) and Aleph (Srinivasan 2007). RSD features satisfy the connectivity requirement, which imposes that no feature can be decomposed into a conjunction of two or more features.

HiFi (Kuželka and Železný 2008)   is a propositionalization approach that constructs first-order features with hierarchical structure. Due to this feature property, the algorithm performs the transformation in a polynomial time of the maximum feature length. The resulting features are the shortest in their semantic equivalence class. The algorithm performs much faster than RSD for longer features.

RelF (Kuželka and Železný 2011)   constructs a set of tree-like relational features by combining smaller conjunctive blocks. The algorithm scales better than other state-of-the-art propositionalization algorithms.

Cardinalization (Ahmed et al. 2015)   is designed to handle not only categorical but also numerical attributes in propositionalization. It handles a threshold on numeric attribute values and a threshold on the number of objects simultaneously satisfying the attribute's condition. Cardinalization can be seen as an implicit form of discretization.

CARAF (Charnay et al. 2015)   approaches the problem of large relational feature search space by aggregates base features into complex compounds, similarly to Relaggs. While Relaggs tackles the overfitting problem by restricting itself to relatively simple aggregates, CARAF incorporates more complex aggregates into a random forest, which ameliorates the overfitting effect.

Aleph (Srinivasan 2007)   is actually an ILP toolkit with many modes of functionality: learning of theories, feature construction, incremental learning, etc. Aleph uses mode declarations to define syntactic bias. Input relations are Prolog clauses, defined either extensionally or intensionally. Aleph's feature construction functionality also means that it is a propositionalization approach.

Wordification (Perovšek et al. 2013, 2015)   is a propositionalization method inspired by text mining that can be viewed as a transformation of a relational database into a corpus of text documents. The distinguishing properties of Wordification are its efficiency when used on large relational datasets and the potential for using text mining approaches on the transformed propositional data. This approach is described in more detail in the following section.

4.5 Wordification: Unfolding Relational Data into BoW Vectors

Inspired by text mining, this section presents a propositionalization approach to relational data mining, named *wordification*. While most other propositionalization techniques first construct complex relational features that act as attributes in the resulting tabular data representation, wordification generates much simpler features with a goal to achieve greater scalability.

4.5.1 Outline of the Wordification Approach

The wordification approach is illustrated in Fig. 4.3. The input to wordification is a relational database, consisting of one main table and several related tables, connected with one-to-one or one-to-many relationships. For each individual (i.e. each row) of the main table (the table on the top of the left part of the figure), one BoW vector d_i is constructed in a two-step process described below.

Feature vector construction.   In the first step, new features are constructed as a combination of table name, name of the attribute, and its discrete (or discretized) value as:

$$tableName\_attributeName\_attributeValue. \qquad (4.1)$$

Such features are referred to as *word-items* in the rest of this section. Note that the values of every non-discrete attribute need to be discretized to be able to represent

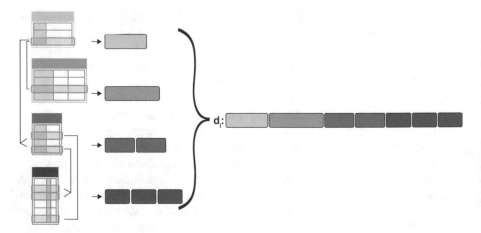

Fig. 4.3 The transformation from a relational database representation into the BoW feature vector representation, illustrated for an individual instance (row) of the main data table

them as word-items. For each individual, the word-items are first generated for the main table and then for each entry from the related tables, and finally joined together according to the relational schema of the database. For each instance of the main data table:

- for one-to-one relationships between tables, we get one itemset of word-items in the output vector,
- for one-to-many relationships we get several itemsets in the output vector.

Concatenation and normalization.   In the second step, for a given data instance, all the constructed itemsets of word-items are concatenated into a joint feature vector d_i. The output of wordification is a corpus D of 'text documents', consisting of BoW vectors d_i constructed for each instance of the main data table. In every d_i, individual word-items are weighted with TF-IDF weights, computed using Eq. (3.1) on page 60: a word-item with a high TF-IDF value will be considered important for the given document d_i if it is frequent within document d_i and infrequent in the entire document corpus D. Consequently, the weight of a word-item indicates how relevant this feature is for the given individual. If appropriate, TF-IDF weights can then be used to filter out word-items with low importance or use them directly by a propositional learner.

In the rest of this section, we refer to individuals as documents, features as word-items, and the resulting representation as the BoW representation. The wordification approach is illustrated on a modified version of the East-West trains dataset, which has been largely simplified for clarity the presentation of the wordification approach. In the simplified dataset shown in Table 4.5, we consider only one eastbound and one westbound train, each with just two cars with certain properties. In this problem, the TRAIN table is the main table, where the trains are the instances, and the task is to learn a classifier to determine the direction of an unseen train. As mentioned, the idea of wordification is transform an individual (a row in the table) to a 'document', formed as an itemset, consisting of word-items. The 'documents' representing trains t1 and t5 are shown in Table 4.6.

To overcome the loss information (resulting from building a 'document' for each instance (each row in the main table) by concatenating all word-items from multiple instances (rows) of the connected tables into a single document), n-grams of word-items, constructed as conjunctions of several word-items, can also be considered.

Table 4.5 Simplified East-West trains relational database representation of the dataset

TRAIN		CAR				
trainID	Class	carID	shape	roof	wheels	trainID
t1	eastbound	c11	rectangle	none	2	t1
t5	westbound	c12	rectangle	peaked	3	t1
		c51	rectangle	none	2	t5
		c52	hexagon	flat	2	t5

Table 4.6 The BoW representation of East-West trains dataset using the wordification approach, using simple unigrams of word-items

```
t1: [car_roof_none, car_shape_rectangle, car_wheels_2, car_roof_peaked,
     car_shape_rectangle, car_wheels_3], eastbound
t5: [car_roof_none, car_shape_rectangle, car_wheels_2, car_shape_hexagon,
     car_wheels_2], westbound
```

Technically, n-gram construction takes every combination of k-word-items ($1 \leq k \leq n$) from the set of all word-items corresponding to the given individual. The k-grams are constructed as follows:

$$\text{word-item}_1\_\_\text{word-item}_2\_\_ \ldots \_\_\text{word-item}_k,$$

where each word-item is a combination of the table name, name of the attribute, and its discrete value. The instances are concatenated using the '__' concatenation symbol.

Take again the example, where for two trains t1 and t5, the corresponding 'documents' (one for each train) are generated, as shown in the top part of Table 4.7. Most frequently, and as is the case in the example in Table 4.7, just bigrams of two word-items are used. After this, the documents are transformed into the BoW representation as shown in the middle part of Table 4.7, followed by calculating the TF-IDF values for each word of each document. Finally, the class attribute column is appended to the transformed BoW table, as shown in the bottom part of Table 4.7.

For a fixed vocabulary, the constructed set of 'documents' can be represented in the form of a data table, consisting of rows representing individual data instances, columns representing individual word-items (or k-grams of word-items), and fields representing their TF-IDF weights. After this transformation, traditional machine learning methods can be employed on the transformed data table.

4.5.2   Wordification Algorithm

The wordification algorithm consists of two transformation steps, implemented with functions wordification and wordify, with their pseudocode shown in Fig. 4.4.

The algorithm starts with recursive document construction on the main table instances (lines 4–9 of the wordification algorithm). It first creates word-items for the attributes of the target table (lines 3–7 of the wordify algorithm), followed by concatenations of the word-items and results of the recursive search through examples of the connecting tables (lines 9–17 of the wordify algorithm). This document construction step is done independently for each example of the main table. This allows simultaneous search along the tree of connected tables. Lines 4–9 of the wordification algorithm can be run in parallel, which allows for a significant speed-up. A common obstacle in parallel computing is memory synchronization between different subtasks, which is not problematic here, as concurrent processes

Table 4.7 The BoW representation of East-West trains dataset using the wordification approach.
It first presents the two trains as sets of features consisting of unigrams and bigrams of word-items.
It then presents two tabular representations of the dataset, the first using counts and the second
using TF-IDF weights of features

```
t1: [car_roof_none, car_shape_rectangle, car_wheels_2,
     car_roof_none__car_shape_rectangle, car_roof_none__car_wheels_2,
     car_shape_rectangle__car_wheels_2, car_roof_peaked, car_shape_rectangle,
     car_wheels_3, car_roof_peaked__car_shape_rectangle,
     car_roof_peaked__car_wheels_3, car_shape_rectangle__car_wheels_3],
     eastbound
t5: [car_roof_none, car_shape_rectangle, car_wheels_2,
     car_roof_none__car_shape_rectangle, car_roof_none__car_wheels_2,
     car_shape_rectangle__car_wheels_2, car_roof_flat, car_shape_hexagon,
     car_wheels_2, car_roof_flat__car_shape_hexagon,
     car_roof_flat__car_wheels_2, car_shape_hexagon__car_wheels_2],
     westbound
```

ID	car_shape _rectangle	car_roof _peaked	car_wheels_3	car_roof_peaked__ car_shape_rectangle	car_shape_rectangle __car_wheels_3	...	Class
t1	2	1	1	1	1	...	eastbound
t5	1	0	0	0	0	...	westbound

ID	car_shape _rectangle	car_roof _peaked	car_wheels_3	car_roof_peaked__ car_shape_rectangle	car_shape_rectangle __car_wheels_3	...	Class
t1	0.000	0.693	0.693	0.693	0.693	...	eastbound
t5	0.000	0.000	0.000	0.000	0.000	...	westbound

only need to share a cached list of subtrees. This list stores the results of subtree
word concatenations in order to visit every subtree only once.

As wordification can produce a large number of features (word-items), especially
when the maximal number of n-grams per word-items is large, we perform pruning
of words that occur in less than a predefined percentage (5% on default) of
documents. This reduces trees' size by removing their branches that are expected
to provide little power for instance classification.

The constructed features are simple. As we do not explicitly use existential
variables in the new features (word-items), we instead rely on the TF-IDF measure
to implicitly capture the importance of a word-item for a given individual. In the
context of text mining, the TF-IDF value reflects how representative is a certain
feature (word-item) for a given individual (document).

4.5.3 Improved Efficiency of Wordification Algorithm

This section presents an implementation of the wordification algorithm, developed
with the aim to improve its *scalability* (Lavrač et al. 2020), which was achieved by
the following design decisions.

```
1 wordification (T, p, k)
   Input           : target table T, pruning percentage p, maximal number of word-items per
                     word k
   Output          : propositionalized table R with TF-IDF values, corpus of documents D
2 D ← []
3 W ← / Φ                                               // vocabulary set
4 for ex ∈ T do
5                                          // entries of the target table
6     d ←wordify (T,ex,k)                      // construct the document
7     D ← D + [d]                       // append document to the corpus
8     W ← W ∪ keys(d)                        // extend the vocabulary
9 end
10 W ←prune (W,p)                                  // optional step
11 return [ calculateTFIDFs (D,W),D]
```

```
1 wordify (T, ex, k)
   Input           : table T, example ex from table T, maximal number of word-items per
                     word k
   Output          : document collection d
2 d ← {}                    // hash table with the default size of 0
3 for i ← 1 to k do                          // for all word-item lengths
4     for comb ∈ attrCombs (ex,k) do // attr. combinations of length k
5         d[word(comb)] ← d[word(comb)] + 1
6     end
7 end
8 // for every connected table through an example
9 for secTable ∈ connectedTables (T) do
10    for secEx ∈ secTable do
11        if primaryKeyValue (ex)=foreignKeyValue (secEx) then
12            for (word,count) ∈ wordify (secTable,secEx,k) do
13                d[word] ← d[word] + count
14            end
15        end
16    end
17 end
18 return d
```

Fig. 4.4 Pseudocode of wordification, followed by the pseudocode of creation of one document in function wordify

- The input no longer needs to be hosted in a relational database. The algorithm supports SQL conventions, as commonly used in the ILP community.[1] This modification renders the method completely local, enabling offline execution

---

[1] https://relational.fit.cvut.cz/.

without additional overhead. Such a setting also offers easier parallelism across computing clusters.

- The algorithm is implemented in Python 3 with minimum dependencies for computationally more intense parts, such as the Scikit-learn (Pedregosa et al. 2011), Pandas, and Numpy libraries (Van Der Walt et al. 2011). All database operations are implemented as array queries, filters, or similar.
- As shown by Perovšek et al. (2015), wordification's caveat is an extensive sampling of (all) tables. We relax this constraint to close (up to second order) foreign key neighborhood, notably speeding up the relational item sampling part, but with some loss in relational item diversity. Minimal relational item frequency can be specified for larger databases, constraining potentially noisy parts of the feature space.

One of the most apparent problems of the original wordification algorithm is its spatial complexity. This issue is addressed as follows:

- Relational items are hashed for minimal spatial overhead during sampling.
- During the construction of the final representation, a sparse matrix is filled based on relational item occurrence.
- The matrix is serialized directly into list-like structures.
- Only the final representation is stored as a low-dimensional (e.g., 32) dense matrix.

This implementation of wordification is used in the two pipelines unifying propositionalization and embeddings, presented in Chap. 6, where we present two unifying methods, PropStar and PropDRM, which combine propositionalization and embeddings by capturing relational information through propositionalization and then applying deep neural networks to obtain its dense embeddings. In the propositionalization step, both methods use the implementation of the wordification algorithm presented in this section.

4.6   Deep Relational Machines

Training deep neural networks on propositionalized relational data was explored by Srinivasan et al. (2019), following the work on Deep Relational Machines (DRMs) introduced by Lodhi (2013). In Lodhi's work, where DRMs were used to predict protein folding properties and the mutagenicity assessment of small molecules, the DRMs used bodies of Prolog sentences (first-order Horn clauses) as inputs to restricted Boltzmann machines. As an example of DRM inputs, consider a simplified propositional representation of five instances (rows) in Table 4.8, where complex features are bodies of Horn clauses, formed of conjuncts of features f_i.

The initial studies using DRMs explored how deep neural networks can be used to extend relational learning. Lodhi (2013) addressed the problem that propositionalized datasets, represented as sparse matrices, are a challenge for

Table 4.8 An example input
to a deep relational machine
that operates on the instance
level

Instance	$f_1 \wedge f_2$	$f_3 \wedge f_2$	$f_1 \wedge f_3$	$f_5 \wedge f_2$	$f_4 \wedge f_1 \wedge f_5$	Class
1	1	1	1	1	0	+
2	0	1	0	0	1	+
3	0	0	1	0	0	−
4	0	1	0	0	1	−
5	1	0	0	0	1	−

deep neural networks, which are effective for learning from numeric data by constructing intermediate knowledge concepts improving the semantics of baseline input representations. Since sparse matrices resulting from propositionalization are not a suitable input to deep neural networks, the DRM approach by Lodhi (2013) used feature selection with information-theoretic measures, such as information gain, as means to address this challenge.

In summary, DRMs address the following issues at the intersection of deep learning and relational learning:

- DRMs demonstrated that deep learning on propositionalized relational structures is a sensible approach to relational learning.
- Their input is comprised of logical conjuncts, offering the opportunity to obtain human-understandable explanations.
- DRMs were successfully employed for classification and regression.
- Ideas from the area of representation learning have only recently been explored in the relational context (Dumančić et al. 2018), indicating there are possible improvements both in terms of execution speed, as well as more informative feature construction on the symbolic level.

Recently, promising results were demonstrated in the domain of molecule classification (Dash et al. 2018) using the ILP learner Aleph in its propositionalization mode for feature construction. After obtaining the propositional representation of data, the obtained data table was fed into a neural network to predict the target class (e.g., a molecule's activity). Again, the sparsity and size of the propositionalized representation was a problem for deep neural networks. Concerning the interpretability of DRMs, the work of Srinivasan et al. (2019) proposes a logical approximation of the LIME explanation method (Ribeiro et al. 2016).

The development of DRMs that are efficient concerning both space and time is an ongoing research effort. Building on the ideas of DRMs, a variant of this approach, named PropDRM (Lavrač et al. 2020) is capable of learning directly from large, sparse matrices that are returned from wordification of a given relational database, rather than using feature selection or the output of Aleph's feature construction approach. An efficient implementation of PropDRM is presented in Sect. 6.2.2.

4.7 Implementation and Reuse

This section demonstrates the wordification algorithm and introduces the *python-rdm* package, which simplifies relational data mining by integrating wrappers for several relational learning algorithms.

4.7.1 Wordification

The wordification approach, introduced in Sect. 4.5, is demonstrated on the East-West trains dataset (Michie et al. 1994), provided in the CSV (Comma Separated Values) format, where files contain additional headers defining relations between tables (i.e. their primary and foreign keys) as well as attribute types. We demonstrate different settings of the wordification algorithm. The resulting features are ranked, selected, and used with a decision tree learner to build a classifier. The Jupyter notebook with code is available in the repository of this monograph: https://github.com/vpodpecan/representation_learning/blob/master/Chapter4/wordification.ipynb.

4.7.2 Python-rdm Package

While several relational learning and relational data mining tools are open source, they are not easily accessible for machine learning practitioners. The *python-rdm*[2] package is a collection of wrappers for relational data mining (RDM) algorithm implementations that aims to remedy this problem. As input data, the package supports relational databases such as MySQL, PostgreSQL, and SQLite, as well as plain CSV text files. The following RDM algorithms are available: Aleph, RSD, Wordification, RelF, HiFi, Caraf, Relaggs, 1BC, and Tertius. The complete documentation of *python-rdm* is available online.[3]

The Jupyter notebook that demonstrates python-rdm consists of two parts. The first part demonstrates the TreeLiker software on the East-West trains dataset (Michie et al. 1994), which is read from a remote MySQL database. Once the data is read and converted into the appropriate format, the RelF algorithm is run to induce features. Then, feature selection is applied and a decision tree is built and visualized using the scikit-learn library (Pedregosa et al. 2011). In the second part, the RSD algorithm, briefly described in Sects. 4.3.2 and 4.4, is applied on the Mutagenesis dataset (Srinivasan et al. 1994). First, continuous attributes are discretized using equal frequency discretization. Then, RSD is applied to induce

---

[2]https://pypi.org/project/python-rdm/.
[3]https://python-rdm.readthedocs.io/en/latest/.

features of maximal length 4. The resulting table is split into train and test data, a decision tree classifier is trained and evaluated, and the tree's top features are shown. The Jupyter notebook is available in the repository of this monograph: https://github. com/vpodpecan/representation_learning/blob/master/Chapter4/python-rdm.ipynb.

References

Chowdhury Farhan Ahmed, Nicolas Lachiche, Clément Charnay, Soufiane El Jelali, and Agnès Braud. Flexible propositionalization of continuous attributes in relational data mining. *Expert Systems with Applications*, 42(21):7698–7709, 2015.

Ivan Bratko. *Prolog Programming for Artificial Intelligence*. Addison-Wesley, Wokingham, 2nd edition, 1990.

Clément Charnay, Nicolas Lachiche, and Agnès Braud. CARAF: Complex aggregates within random forests. In *Proceedings of the 25th International Conference on Inductive logic programming*, pages 15–29, 2015.

Peter Clark and Tim Niblett. The CN2 induction algorithm. *Machine Learning*, 3(4):261–283, 1989.

Tirtharaj Dash, Ashwin Srinivasan, Lovekesh Vig, Oghenejokpeme I Orhobor, and Ross D King. Large-scale assessment of deep relational machines. In *Proceedings of the International Conference on Inductive Logic Programming*, pages 22–37, 2018.

Sebastijan Dumančić, Tias Guns, Wannes Meert, and Hendrik Blockleel. Auto-encoding logic programs. In *Proceedings of the International Conference on Machine Learning*, 2018.

Sašo Džeroski and Nada Lavrač, editors. *Relational Data Mining*. Springer, Berlin, 2001.

Peter Flach and Nicholas Lachiche. 1BC: A first-order Bayesian classifier. In *Proceedings of the 9th International Workshop on Inductive Logic Programming (ILP-99)*, pages 92–103. Springer, 1999.

Peter Flach and Nicholas Lachiche. Confirmation-guided discovery of first-order rules with Tertius. *Machine Learning*, 42(1/2):61–95, 2001.

Lise Getoor. *Introduction to Statistical Relational Learning*. The MIT Press, 2007.

Arno J. Knobbe. *Multi-Relational Data Mining*, volume 145. IOS Press, 2005.

Stefan Kramer, Bernhard Pfahringer, and Christoph Helma. Stochastic propositionalization of non-determinate background knowledge. In *Proceedings of the 8th International Conference on Inductive Logic Programming (ILP-2000)*, pages 80–94, 2000.

Stefan Kramer, Nada Lavrač, and Peter Flach. Propositionalization approaches to relational data mining. In Sašo Džeroski and Nada Lavrač, editors, *Relational Data Mining*, pages 262–291. Springer, 2001.

Mark A. Krogel and Stefan Wrobel. Transformation-based learning using multirelational aggregation. In *Proceedings of International Conference on Inductive Logic Programming*, pages 142–155. Springer, 2001.

Mark A. Krogel, Simon Rawles, Filip Železný, Peter Flach, Nada Lavrač, and Stefan Wrobel. Comparative evaluation of approaches to propositionalization. In *Proceedings of the 13th International Conference on Inductive Logic Programming (ILP-2003)*, pages 197–214, 2003.

Ondřej Kuželka and Filip Železný. Block-wise construction of tree-like relational features with monotone reducibility and redundancy. *Machine Learning*, 83(2):163–192, 2011.

Ondřej Kuželka and Filip Železný. HiFi: Tractable propositionalization through hierarchical feature construction. In *Late Breaking Papers, the 18th International Conference on Inductive Logic Programming*, pages 69–74, 2008.

Nicolas Lachiche and Peter Flach. 1BC2: A true first-order Bayesian classifier. In *Proceedings of Inductive Logic Programming*, pages 133–148, 2003.

Nada Lavrač and Sašo Džeroski. *Inductive Logic Programming: Techniques and Applications.* Ellis Horwood, 1994.

Nada Lavrač and Peter Flach. An extended transformation approach to inductive logic programming. *ACM Transactions on Computational Logic*, 2(4):458–494, 2001.

Nada Lavrač, Sašo Džeroski, and Marko Grobelnik. Learning nonrecursive definitions of relations with LINUS. In *Proceedings of the 5th European Working Session on Learning (EWSL-91)*, pages 265–281, 1991.

Nada Lavrač, Blaž Škrlj, and Marko Robnik-Šikonja. Propositionalization and embeddings: Two sides of the same coin. *Machine Learning*, 109:1465–1507, 2020.

Huma Lodhi. Deep Relational Machines. In *Proceedings of the International Conference on Neural Information Processing*, pages 212–219, 2013.

Donald Michie, Stephen H. Muggleton, David Page, and Ashwin Srinivasan. To the international computing community: A new East-West challenge. Technical report, Oxford University Computing laboratory, 1994.

Stephen H. Muggleton, editor. *Inductive Logic Programming.* Academic Press, London, 1992.

Stephen H. Muggleton. Inverse entailment and Progol. *New Generation Computing*, 13(3–4): 245–286, 1995.

Fabian Pedregosa, Gaël Varoquaux, Alexandre Gramfort, Vincent Michel, Bertrand Thirion, Olivier Grisel, Mathieu Blondel, Peter Prettenhofer, Ron Weiss, Vincent Dubourg, Jake Vanderplas, Alexandre Passos, David Cournapeau, Matthieu Brucher, Matthieu Perrot, Edouard Duchesnay, and Gilles Louppe. Scikit-learn: Machine learning in Python. *Journal of Machine Learning Research*, 12:2825–2830, 2011.

Matic Perovšek, Anže Vavpetič, Bojan Cestnik, and Nada Lavrač. A wordification approach to relational data mining. In *Proceedings of the International Conference on Discovery Science*, pages 141–154, 2013.

Matic Perovšek, Anze Vavpetič, Janez Kranjc, Bojan Cestnik, and Nada Lavrač. Wordification: Propositionalization by unfolding relational data into bags of words. *Expert Systems with Applications*, 42(17–18):6442–6456, 2015.

J. Ross Quinlan. Learning logical definitions from relations. *Machine Learning*, 5:239–266, 1990.

Marco Tulio Ribeiro, Sameer Singh, and Carlos Guestrin. Why should I trust you?: Explaining the predictions of any classifier. In *Proceedings of the 22nd ACM SIGKDD International Conference on Knowledge Discovery and Data Mining*, pages 1135–1144. ACM, 2016.

Jorma Rissanen. A universal prior for integers and estimation by minimum description length. *The Annals of Statistics*, 11(2):416–431, 1983.

Ashwin Srinivasan. *The Aleph Manual.* University of Oxford, 2007. Online. Accessed 26 October 2020. https://www.cs.ox.ac.uk/activities/programinduction/Aleph/.

Ashwin Srinivasan, Stephen H. Muggleton, Ross D. King, and Michael J. E. Sternberg. Mutagenesis: ILP experiments in a non-determinate biological domain. In *Proceedings of the 4th International Workshop on Inductive Logic Programming, volume 237 of GMD-Studien*, pages 217–232, 1994.

Ashwin Srinivasan, Lovekesh Vig, and Michael Bain. Logical explanations for Deep Relational Machines using relevance information. *Journal of Machine Learning Research*, 20(130):1–47, 2019.

Stefan Van Der Walt, Chris Colbert, and Gaël Varoquaux. The NumPy array: A structure for efficient numerical computation. *Computing in Science & Engineering*, 13(2):22, 2011.

Filip Železný and Nada Lavrač. Propositionalization-based relational subgroup discovery with RSD. *Machine Learning*, 62:33–63, 2006.

Chapter 5
Graph and Heterogeneous Network Transformations

This chapter presents the key ideas related to embedding simple (homogeneous) and more elaborate (heterogeneous) graphs. It first outlines some of the key approaches for constructing embeddings based exclusively on graph topology. We then discuss recent advances that include feature information into the constructed graph embeddings. We also address data transformation methods applicable to graphs with different types of nodes and edges, referred to as heterogeneous graphs or heterogeneous information networks. The chapter is structured as follows. Section 5.1 presents approaches to embedding simple homogeneous graphs, starting with the popular DeepWalk and node2vec methods, followed by a selection of other random walk based graph embedding methods. Section 5.2 introduces heterogeneous information networks, containing nodes of different types, the most useful tasks applied to such networks, and selected approaches to embedding heterogeneous information networks. We present a method for propositionalizing text enriched heterogeneous information networks and a method for heterogeneous network decomposition in Sect. 5.3. Ontology transformations for semantic data mining are presented in Sect. 5.4. Selected techniques for embedding knowledge graphs are presented in Sect. 5.5. The chapter concludes by presenting selected methods implemented in Jupyter Python notebooks in Sect. 5.6.

5.1 Embedding Simple Graphs

Datasets used in data mining and machine learning are usually available in a tabular form, where an *instance* (corresponding to a row in the data table) is characterized by its properties described with values of a set of attributes (each corresponding to a table column). The rows in the table are unrelated to each other. In contrast, the motivation for mining graphs and more general information networks is the fact that information may exist both at the level of instances (i.e. nodes of a graph) as well as

© Springer Nature Switzerland AG 2021
N. Lavrač et al., *Representation Learning*,
https://doi.org/10.1007/978-3-030-68817-2_5

in the way how the instances *interact* (expressed with edges). A way to incorporate such information into a tabular form is via network embedding.

Network embedding is a mechanism for converting networks into a tabular/matrix representation, where each network node is represented as a vector of fixed predefined length d, and a set of nodes is represented as a table with d columns. Such a conversion aims to preserve the structural properties of a network, which is achieved by preserving node similarity in tabular representations, by converting node similarity into vector similarity.

Let $G = (V, E)$ represent a simple graph, where V is the set of nodes and E the set of edges. The process of graph embedding corresponds to a mapping $f : G \rightarrow \mathbb{R}^{|V| \times d}$, i.e. a mapping that takes G as the input and projects it to a $|V| \times d$ dimensional vector space, where each node $u \in V$ gets assigned a d dimensional vector of real numbers. The rationale for the construction of algorithms that approximate f is to learn a representation in the form of a feature matrix, given that many existing data mining approaches operate on feature matrices. The feature matrix can be used in combination with standard machine learners to solve network analysis tasks, e.g., node classification or link prediction task, introduced in Sect. 2.4.1.

The first studies of graph embeddings were significantly influenced by embeddings construction from textual data. For example, the well known skip-gram model, initially used in word2vec (Mikolov et al. 2013) (see Sect. 3.3.1, page 66), was successfully adapted to learn node representations. The skip-gram approach to node embeddings is illustrated in Fig. 5.1. The idea is to use random walks as documents and individual nodes as words. For a given node, nearby nodes in the same random walk establish a context used to construct positive examples in the word2vec approach. The embedding vectors are extracted from the hidden layer of the trained neural network.

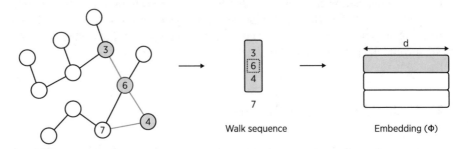

Walk sequence   Embedding (Φ)

Fig. 5.1 Schematic representation of the skip-gram node embeddings. For each node, its representation is obtained based on its neighbors in the graph. The neighborhood is obtained from the random walk. The example shows the construction of an embedding of dimension d for node number 6 that has neighbors 3 and 4 in the random walk (3, 6, 4, 7). If the word2vec-like approach is used to get the embeddings, d corresponds to the number of hidden nodes in the neural network

5.1.1 DeepWalk Algorithm

DeepWalk (Perozzi et al. 2014) is one of the first embedding algorithms that treat nodes as words and short random walks in graphs as sentences to learn node embeddings. It does not only learn a distribution of node cooccurrences in random walks, but it learns individual node representations by maximizing the likelihood of observing neighboring nodes in a set of short random walks given the representation of the node. More formally, let $\Phi(v)$ represent the v-th node's embedding. DeepWalk solves the following optimization problem:

$$\arg\max_{\Phi} \left[\log Pr(\{v_{i-w}, \ldots v_{i-1}, v_{i+1}, \ldots v_{i+w}\} \,|\, \Phi(v_i)) \right],$$

where $\{v_{i-w}, \ldots v_{i-1}, v_{i+1}, \ldots v_{i+w}\}$ represents a set of nodes appearing in the context of node v_i (without v_i). The context is established with a number of random walks starting in node v_i. Thus, the expression tries to maximize the likelihood of observing a given node's neighborhood with respect to its representation. The resulting representation for graph $G = (V, E)$ is a $\mathbb{R}^{|V| \times d}$ matrix that can be used for plethora of down-stream machine learning tasks on networks, such as node classification, link prediction, graph comparison etc. The most popular variant of this idea, node2vec (Grover and Leskovec 2016), is presented next, followed by some of the other successful variations.

5.1.2 Node2vec Algorithm

The node2vec network embedding algorithm (Grover and Leskovec 2016) uses a random walk approach to calculate features that express similarities between node pairs. For a user-defined number of columns d, the algorithm returns a matrix $F^* \in \mathbb{R}^{|V| \times d}$, defined as follows:

$$F^* = \text{node2vec}(G) = \arg\max_{F \in \mathbb{R}^{|V| \times d}} \sum_{u \in V} \left(-\log(Z_u) + \sum_{n \in N(u)} F(n) \cdot F(u) \right),$$

(5.1)

where $N(u)$ denotes the network neighborhood of node u, $F(n) \cdot F(u)$ denotes the dot product, and Z_u is explained further down. In the above expression, each matrix F is a collection of d-dimensional feature vectors, with the i-th row of the matrix corresponding to the feature vector of the i-th node in the network. We write $F(u)$ to denote the row of matrix F corresponding to node u. The resulting maximal matrix is denoted with F^*.

The goal of node2vec is to construct feature vectors $F(u)$ in such a way that feature vectors of nodes that share a certain neighborhood will be similar. The inner sum $\sum_{n \in N(u)} F(n) \cdot F(u)$, maximized in Eq. 5.1, calculates the similarities between node u and all nodes in its neighborhood. It is large if the feature vectors of nodes in the same neighborhood are colinear; however, it also increases if feature vectors of nodes have a large norm. Value Z_u calculates the similarities between node u and all the nodes in the network as follows:

$$Z_u = \sum_{v \in V} e^{F(u) \cdot F(v)}.$$

Note that value of $-\log(Z_u)$ decreases when the norms of feature vectors $F(v)$ increase, thereby penalizing collections of feature vectors with large norms.

Grover and Leskovec (2016) have shown that Eq. 5.1 has a probabilistic interpretation that models a process of randomly selecting nodes from the network, modeling the probabilities $P(n|u)$ of node n following node u. Neighborhood $N(u)$ in Eq. 5.1 is obtained by simulating a random walker traversing the network starting at node u. Unlike the PageRank random walker (see Sect. 2.4.1, page 38), the transition probabilities for traversing from node n_1 to node n_2 depend on node n_0 that the walker visited before node n_1, making the process of traversing the network a second order random walk. The non-normalized transition probabilities are set using two parameters, p and q, and are equal to:

$$P(n_2 | \text{moved from } n_0 \text{ to } n_1 \text{ in previous step}) = \begin{cases} \frac{1}{p} & \text{if } n_2 = n_0 \\ 1 & \text{if } n_2 \text{ can be reached from } n_1 \\ \frac{1}{q} & \text{otherwise} \end{cases}$$

The parameters of the expression are referred to as the *return* parameter p and the *in-out* parameter q. A low value of the return parameter p means that the random walker is more likely to backtrack its steps and the random walk will be closer to a breadth first search. On the other hand, a low value of parameter q encourages the walker to move away from the starting node and the random walk resembles a depth first search of the network. To calculate the maximizing matrix F^*, a set of random walks of limited size is simulated starting from each node in the network to generate samples of sets $N(u)$.

The function maximizing Eq. 5.1 is calculated using the stochastic gradient descent. The matrix of feature vectors node2vec(G) in Eq. 5.1 is estimated at each sampling of neighborhoods $N(u)$ for all nodes in the network. The resulting matrix F^* maximizes the expression for the simulated neighborhood set.

5.1.3 Other Random Walk-Based Graph Embedding Algorithms

We present some other successful approaches to graph embeddings involving random walk based neighborhoods.

LINE. The LINE algorithm (Tang et al. 2015b) extends the idea of DeepWalk and node2vec. Node2vec takes into account different types of random walks, parameterized by the breadth and depth-first search. For node classification and link prediction tasks, LINE performs similarly well by taking into account the local and global network structure but can efficiently handle large networks.

PTE. The PTE algorithm (Tang et al. 2015a) exploits heterogeneous networks with texts for supervised embedding construction—they leverage the document labels to improve the quality of the final embeddings.

NetMF. The NetMF algorithm (Qiu et al. 2018), a generalization of DeepWalk, node2vec, LINE, and PTE, re-formulates these algorithms as a matrix factorization problem, indicating that such type of node embedding algorithms can be described as part of a more general theoretical framework.

struc2vec. The struc2vec approach (Ribeiro et al. 2017) builds on two main ideas: representations of two nodes must be close if the two nodes are structurally similar, and the latent node representation should not depend on any node or edge attribute, including the node labels—the structural identity of a given pair of nodes must be independent of their 'position' in the network. Varying neighborhood size, struc2vec attempts to determine node similarity, useful for learning node representations.

HARP. The HARP algorithm (Chen et al. 2018) for hierarchical representation learning for networks addresses a weakness of DeepWalk and node2vec, which randomly initialize node embeddings during the optimization of embeddings. As these two methods optimize a non-convex objective function, this can lead to local optima. HARP avoids local optima by creating a hierarchy of nodes. It aggregates nodes in each layer of the hierarchy using graph coarsening. The embeddings of higher-level graphs are used to initialize lower levels, all the way down to the original full graph. HARP can be used in conjunction with random walk based methods like DeepWalk and node2vec to obtain better solutions.

There are many other variants of the above methods. For further information on this topic, we refer the reader to recent survey papers of Goyal and Ferrara (2018) and Chen et al. (2020).

5.2 Embedding Heterogeneous Information Networks

In Sect. 5.1, we discussed some of the well-known approaches for obtaining structural embeddings of simple graphs (graphs with a single type of nodes and

Embedding (Φ)

Fig. 5.2 Schematic representation of embedding a node in a heterogeneous network. Nodes in such networks can be of different types (denoted with triangles, circles and rectangles)

edges, referred to as homogeneous networks). These embeddings serve as latent representations of the given graph's structural properties, such as communities and dense sub-graphs. However, many real-world graphs consist of multiple node or edge types. *Heterogeneous information networks* describe heterogeneous types of entities and different types of relations. Embeddings that operate on heterogeneous graphs are schematically illustrated in Fig. 5.2.

This section starts by introducing heterogeneous information networks, followed by examples of such networks and an outline of selected methods for embedding heterogeneous information networks, focusing of graph-convolutional neural networks.

5.2.1   Heterogeneous Information Networks

An information network is a network composed of entities (for example, web pages or research papers) that are in some way connected to other entities (one page may contain links to other pages). As defined in Sect. 5.1, such structures are represented by graphs. However, while graphs are a convenient way to represent relations between different entities, they do not contain any real data. Therefore we introduce a different structure, an *information network*, which refers to a graph in which each node has certain properties. Information networks are a richer way of representing data than either graphs or tables but can lack the power to represent truly complex interactions between entities of different types. Following the work of Sun and Han (2012), to cover such complex cases, we introduce the concept of heterogeneous information networks.

Definition 5.1 (Heterogeneous Information Network)   A heterogeneous information network is a tuple $(V, E, \mathscr{A}, \mathscr{E}, \tau, \phi)$, where $G = (V, E)$ is a directed graph, $\mathscr{A}$ a set of object types, and $\mathscr{E}$ a set of edge types. Functions $\tau : V \to \mathscr{A}$ and $\phi : E \to \mathscr{E}$ establish relations between different types of nodes and edges by the following conditions: if edges $e_1 = (x_1, y_1)$ and $e_2 = (x_2, y_2)$ belong to the same edge type (i.e. $\phi(e_1) = \phi(e_2)$), then their start points and their end points belong to the same node type (i.e. $\tau(x_1) = \tau(x_2)$ and $\tau(y_1) = \tau(y_2)$).

Sun and Han (2012) note that sets $\mathscr{A}$ and $\mathscr{E}$, along with the restrictions imposed by the definition of a heterogeneous information network, can be seen as a network as well, with edges connecting two node types if there exists an edge type whose edges connect vertices of the two node types. The authors call this meta-level description of a network a *network schema*, defined below.

Definition 5.2 (Network Schema) A network schema of a heterogeneous information network $G = (V, E, \mathscr{A}, \mathscr{E}, \tau, \phi)$, denoted T_G, is a directed graph with vertices $\mathscr{A}$ and edges $\overline{\mathscr{E}}$, where edge type $t \in \mathscr{E}$ whose edges connect vertices of type $t_1 \in \mathscr{A}$ to vertices of type $t_2 \in \mathscr{A}$, defines an edge in $\overline{\mathscr{E}}$ from type t_1 to t_2.

In Sect. 5.2.2, we give an example of a heterogeneous information network. The network schema of the network, shown in Fig. 5.3, is presented in Fig. 5.6.

5.2.2 Examples of Heterogeneous Information Networks

An information network includes both the network structure and the data attached to individual nodes or edges. Take as an example a heterogeneous information network with different node and edge types, where nodes of one type are text documents. This *text-enriched heterogeneous information network* is a fusion of two different data types: heterogeneous information networks and texts. As an illustrative example, consider a heterogeneous citation network in which text documents are papers (Grčar et al. 2013), shown in Fig. 5.3.

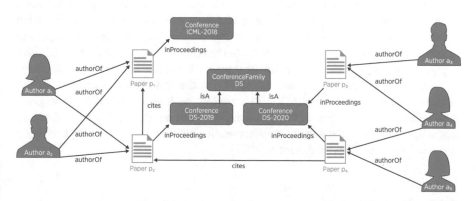

Fig. 5.3 An example of a citation network by Grčar et al. (2013). For brevity, only one direction of each relation is shown (e.g., 'Paper *inProceedings* Conference'), without explicitly presenting its reverse relation (i.e. 'Conference *hasPaper* Paper')

Given a broad definition of heterogeneous information networks, a large amount of human knowledge can be expressed in the form of networks. Some examples are listed below.

Bibliographic networks or citation networks.   These networks connect authors of scientific papers with their papers. Examples of these networks include the DBLP Computer Science Bibliography network examined by Sun and Han (2012) and the text enriched citation network examined by Grčar et al. (2013). In their elementary form, these networks contain at least two types of entities, authors (A) and papers (P), and at least one type of edges, connecting authors to the papers they have (co)authored. The network may also include other entity types, including journals and conferences (which can be merged into one type, *venue*), authors' institutions, and so on. The list of edge types can also be expanded along with the entity types: papers are written by authors and published at venues, and authors are affiliated with institutions. Papers may cite other papers, meaning that a paper in the network can be connected to entities of all other entity types in the network.

Online social networks.   These networks model the structure of popular online social platforms such as Twitter and Facebook. In the case of Twitter, the network entity types are *user, tweet, hashtag*, and *term*. The connections between the types are e.g., users *follow* other users, users *post* tweets, tweets *reply* to other tweets, and tweets *contain* terms (and hashtags).

Biological networks.   These networks denote a large number of different heterogeneous information networks. They can contain various entity types such as species, genes, proteins, metabolites, Gene Ontology (GO) annotations (Ashburner et al. 2000), etc. The types of links between such mixed entities are diverse. For example, genes *belong to* species, genes *encode* proteins, genes *are annotated by* a GO annotation, etc.

Many heterogeneous network analysis tasks exist. In addition to the specialized tasks discussed in Sect. 2.4.2, the tasks and approaches developed for analyzing homogeneous information networks, described in Sect. 2.4.1, can also be adapted to heterogeneous networks. For example, *community detection* (Wang et al. 2017b), which is one of the most common unsupervised approaches, groups the nodes of a network into densely connected sub-networks and enables learning of their unknown properties. Communities in complex biological networks correspond to functionally connected biological entities, such as the proteins involved in cancerogenesis. In social networks, communities may correspond to people sharing common interests (Duch and Arenas 2005).

5.2.3   Embedding Feature-Rich Graphs with GCNs

Real-world graphs often consist of nodes, additionally described by a set of features. *Graph-convolutional neural networks* (GCNs) exploit such additional information

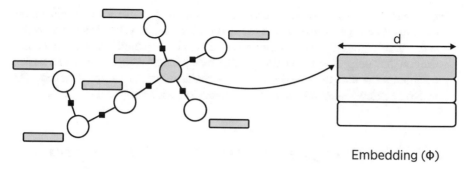

Embedding (Φ)

Fig. 5.4 Schematic representation of a graph-convolutional neural network (GCN). Feature vectors from a target (orange) node's neighbors are aggregated using one of the many possible aggregation functions (e.g., pooling or a separate neural network). Weights (black squares) are shared among all the nodes. The obtained node embeddings contain, apart from structural information, also information from features, yielding better embeddings

when constructing embeddings. Similarly to convolutions over grid-structured input space (such as images and text), graph convolutions also aggregate information over local regions. For example, in images, the convolution considers a pixel's spatial neighborhood and aggregates it with different learned filters. In graphs, the convolution around a node in a graph aggregates the information from its neighborhood.

For a given node, its neighboring nodes' features are *aggregated* into a single representation, which—apart from local topology—contains also feature information from all the considered neighbors. This aggregation starts from the most distant nodes, using predefined aggregation operators, such as max- and mean-pooling. The representation of each node closer to the node of interest is obtained by aggregating its neighbors' feature vectors. This process resembles the convolution operation used in image and text processing neural networks.

A schematic representation of GCNs is shown in Fig. 5.4. Variants of GCNs are mostly used for node classification tasks, even though edges can also be embedded (Kipf and Welling 2016). Below we outline some well-known GCNs.

Spectral GCN.   This neural network architecture (Defferrard et al. 2016) was one of the first GCNs. The authors generalize convolutions over structured grids (e.g., images) to spectral graph theory. They propose fast localized convolutions on graphs, demonstrating that efficient neighborhood-based feature aggregation achieves state-of-the-art classification performance on many datasets.

Laplacian GCN.   This network (Bruna et al. 2014) is similar to the spectral GCN. It demonstrates how the harmonic spectrum of a graph Laplacian can be used to construct neural architectures with a relatively low number of parameters.

GCN for node classification.   This extension of GCNs (Kipf and Welling 2017) is capable of efficient node classification. The authors demonstrate that semi-supervised learning on graphs by using GCNs offers state-of-the-art node classification performance.

GraphSAGE.   This algorithm solves the problem that many node embedding methods do not generalize well to unseen nodes (e.g., when dynamic graphs are considered). GraphSAGE (Hamilton et al. 2017) is an inductive graph embedding learner that outperforms existing alternatives by first sampling a given node's neighborhood and then aggregating the features of the sampled nodes. The method achieves state-of-the-art performance on node classification tasks.

5.2.4   Other Heterogeneous Network Embedding Approaches

Some extensions of the approaches for obtaining simple graph embeddings, presented in Sect. 5.1, are presented below.

metapath2vec.   This algorithm (Dong et al. 2017) leverages a set of pre-defined paths for sampling, yielding embeddings which achieve state-of-the-art performance on node classification tasks. Recently, metapath2vec was successfully applied for drug-gene interactions (Zhu et al. 2018).

OhmNet.   The OhmNet algorithm (Žitnik and Leskovec 2017) is an extension of node2vec to a heterogeneous biological setting, where tissue information is taken into account for gene function prediction. This method empirically outperformed competing tensor factorization methods.

HIN2Vec.   This algorithm (Fu et al. 2017) works with the probability that there is a meta-path between nodes u and v. HIN2Vec generates positive tuples using homogeneous random walks disregarding the type of nodes and links. For each positive instance, it generates several negative instances by replacing node u with a random node.

Heterogeneous data embeddings.   Chang et al. (2015) explored how images, videos, and text could all form a heterogeneous graph, which—when embedded—offers state-of-the-art classification (annotation) performance.

There are many other variants of embedding methods for heterogeneous graph representation learning. Further information on this topic can be obtained in one of the recent survey papers, including the papers by Yang et al. (2020), Dong et al. (2020), and Wang et al. (2020).

5.3   Propositionalizing Heterogeneous Information Networks

This section presents selected methods for propositionalizing heterogeneous information networks. In Sect. 5.3.1, we focus on the TEHmINe method for propositionalizing text-enriched heterogeneous information networks. HINMINE, an advanced approach to heterogeneous information network decomposition, is presented in Sect. 5.3.2.

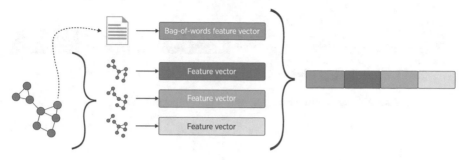

Fig. 5.5 The TEHmINe methodology is performed in three main steps: network decomposition, feature vector construction, and data fusion

5.3.1   TEHmINe Propositionalization of Text-Enriched Networks

The TEHmINe methodology for propositionalizing text-enriched information networks (Grčar et al. 2013) combines elements from text mining and network analysis as a basis for constructing feature vectors describing both the position of nodes in the network and the internal properties of network nodes.

The TEHmINe methodology is performed in three main steps: network decomposition, feature vector construction, and data fusion. The overall three-step methodology is illustrated in Fig. 5.5. Note that an advanced variant of the TEHmINe heterogeneous network decomposition, named HINMINE, is presented in Sect. 5.3.2.

5.3.1.1   Heterogeneous Network Decomposition

The first step of the methodology, i.e. heterogeneous network decomposition, focuses on the network structure. The original heterogeneous information network is decomposed into a set of homogeneous networks. Each homogeneous network is constructed from a circular walk in the original network schema. If a sequence of node types $t_1, t_2, \ldots, t_n$ forms a circular walk in the network schema (meaning that $t_1 = t_n$), then two nodes a and b are connected in the decomposed network if there exists a walk $v_1, v_2, \ldots v_m$, such that $v_1 = a$, $v_m = b$, and each node v_j in the walk is of the same type.

We take the network of scientific papers, shown in Fig. 5.3, to illustrate the network decomposition step of the TEHmINe methodology. The schema of a network, presented in Fig. 5.6, contains entities such as authors, papers, conferences, and conference families. Note that in a directed heterogeneous network, an edge from node v to node w (for example, an author *writes* a paper) implicitly defines an inverse edge going from node w to node v (a paper is *written by* an author).

To construct homogeneous networks with nodes of a single type, we construct relationships between nodes of a given type. Instead of the original directed edges

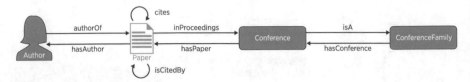

Fig. 5.6 The network schema of the citation network, shown in Fig. 5.3

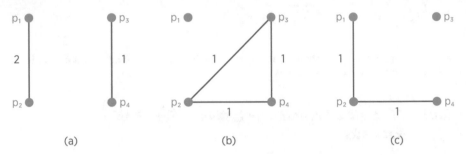

Fig. 5.7 The decomposition of the network from Fig. 5.3 according to the (**a**) Paper–Author–Paper, (**b**) Paper–ConferenceFamily–Paper and (**c**) Paper–Paper paths in the network schema from Fig. 5.6

shown in Fig. 5.3, we construct undirected edges, where each edge corresponds to one circular path in the network schema presented in Fig. 5.6. We label the edges by counting the number of paths in the original network that correspond to the newly constructed undirected edge.

As an illustration, we construct three homogeneous networks of papers, shown in Fig. 5.7. In this figure, the nodes in the decompositions correspond to the papers in the original network. The weights assigned to the edges are the number of paths in the original network. For example, the weights in the Paper-Author-Paper decomposition in Fig. 5.7a, corresponds to the number of authors shared by two papers.

- The first network (Fig. 5.7a) is constructed using the circular walk Paper *hasAuthor* Author *authorOf* Paper in the schema on Fig. 5.6. The resulting relationship between two papers denotes that two papers are connected if they share a common author.
- The second network (Fig. 5.7b) is constructed using two circular walks: (i) Paper *inProceedings* Conference *hasPaper* Paper, and (ii) Paper *inProceedings* Conference *isA* ConferenceFamily *hasConference* Conference *hasPaper* Paper. As a result, two papers are connected if they appear at the same conference or in the same conference family.
- The third network (Fig. 5.7c) is constructed using two circular walks: Paper *cites* Paper or Paper *isCitedBy* Paper. Consequently, two papers are connected if one paper cites another.

This step of the methodology is not necessarily fully automatic, as different meta-paths can be considered for each heterogeneous network. Usually, meta-paths of heterogeneous networks have a real-world meaning, therefore expert judgment should best be used to assess which meta-paths should be defined.

5.3.1.2 Feature Vector Construction

In the second step of the TEHmINe methodology, two types of feature vectors are constructed for each node of the original heterogeneous network. The two types are described below.

BoW vectors. The first type of feature vectors are bag-of-words vectors (see Sect. 3.2.1), which are constructed from the text documents enriching the individual network nodes (e.g., scientific papers). When constructing BoW vectors, each node enriching text is processed using standard text preprocessing techniques. These typically include tokenization, stop-word removal, lemmatization, construction of n-grams of a certain length, and removal of infrequent words from the vocabulary. Using the Euclidean norm, the resulting vectors are normalized.

P-PR vectors. The second type of feature vectors are weighted P-PR vectors obtained by running the Personalized PageRank (P-PR) algorithm (Page et al. 1999) for each node of individual homogeneous networks obtained through network decomposition (see Sect. 5.3.1.1). P-PR vector construction is described below.

For a given node v of a given homogeneous network, the P-PR vector of node v (denoted as P-PR$_v$) contains—for each other node w of the network—the likelihood of reaching w with a random walk starting from v. It is defined as the stationary distribution of the position of a random walker, which starts its walk in node v and (i) either selects one of the outgoing connections or (ii) jumps back to its starting location. Probability p of continuing the walk is a parameter of the algorithm and is set by default to 0.85. The resulting P-PR vectors are normalized according to the Euclidean norm to make them compatible with the BoW vector calculated for the document in the same node.

While there are different ways of computing the P-PR vectors, we describe the iterative version. In the first step, the weight of node v is set to 1, and the other weights are set to 0 to construct r^0, the initial estimation of the P-PR vector. Then, at each step, the weights are spread along the connections of the network using the formula:

$$r^{(k+1)} = p \cdot (A^T \cdot r^{(k)}) + (1 - p) \cdot r^{(0)}, \tag{5.2}$$

where $r^{(k)}$ contains weights of the P-PR vector after k iterations and A is the network's adjacency matrix, normalized so that the elements in each row sum to 1. If all the elements in a given row of the coincidence matrix are zero (i.e. if a node has no outgoing connections), all the values in this row are set to $\frac{1}{n}$, where n is the number of vertices (this simulates the behavior of the walker jumping from a node with no outgoing connections to any other node in the network).

5.3.1.3 Data Fusion

For each network node v, the result of running both the BoW and the P-PR vector construction is a set of vectors $\{v_0, v_1, \ldots, v_n\}$, where v_0 is the BoW vector for the text in node v, and for each i ($1 \leq i \leq n$, where n is the number of network decompositions), v_i is the P-PR vector of node v in the i-th homogeneous network. In the final step of the methodology, these vectors are concatenated to create a concatenated vector v_c for node v:

$$v_c = \sqrt{\alpha_0} v_0 \oplus \sqrt{\alpha_1} v_1 \oplus \cdots \oplus \sqrt{\alpha_n} v_n,$$

constructed by using positive weights $\alpha_0, \alpha_1, \ldots, \alpha_n$ (which sum to 1), where symbol $\oplus$ represents the concatenation of two vectors.

The values of weights α_i are determined by optimizing the weights for a given task, such as node classification. If the multiple kernel learning (MKL) (Rakotomamonjy et al. 2008) optimization algorithm is used, which views the feature vectors as linear kernels, each of the vectors v_i corresponds to a linear mapping $\overline{v_i} : x \mapsto x \cdot v_i$, and the final concatenated vector v_c for node v represents the linear mapping

$$[x_0, x_1, \ldots, x_n] \mapsto \alpha_0 x_0 \cdot v_0 + \alpha_1 x_1 \cdot v_1 + \cdots + \alpha_n x_n \cdot v_n.$$

Another possibility is to fine-tune the weights using a general purpose optimization algorithm, such as differential evolution (Storn and Price 1997).

5.3.2 HINMINE Heterogeneous Networks Decomposition

The HINMINE methodology (Kralj et al. 2018), which is designed for propositionalization of heterogeneous networks that do not include texts enriching the individual network nodes, is motivated by a simplified example shown in Fig. 5.8.

In the network in Fig. 5.8, two papers are connected if they share a common author. In the network decomposition step of the TEHmINe methodology, explained in Sect. 5.3.1.1, the weight of a link between two base nodes (papers) is equal to the number of intermediate nodes (authors) that are linked to (are authors of) both base nodes (papers). Formally, the weight of a link between nodes v and u is

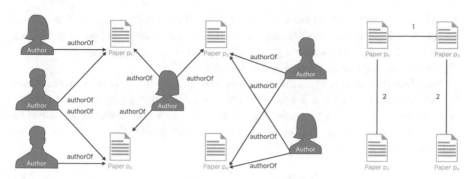

Fig. 5.8 A simplified example of a heterogeneous network in Fig. 5.3 (on the left-hand side) and a homogeneous network extracted from it (on the right-hand side). In the homogeneous network, papers are connected if they share a common author. Weights of the edges are equal to the number of authors that contributed to both papers

$$w(v, u) = \sum_{\substack{m \in M \\ m \text{ is linked to } v \text{ and } u}} 1, \tag{5.3}$$

where M is the set of all intermediate nodes.

A weakness of the TEHmINe approach is that it treats all the nodes equally, which may not be appropriate from the information content point of view. For example, if two papers share an author who only co-authored a small number of papers, it is more likely that these two papers are similar than if the two papers share an author that co-authored tens or even hundreds of papers. The first pair of papers should therefore be connected by a stronger weight than the second. Similarly, if papers are labeled by the research field, two papers sharing an author publishing in only one research field are more likely to be similar than if they share an author who has co-authored papers in several research fields. Again, the first pair of papers should be connected by an edge with a larger weight.

The HINMINE methodology (Kralj et al. 2018) resolves the above identified weakness of the TEHmINe methodology. To this end, in the network decomposition step when decomposing a heterogeneous network into simpler homogeneous networks, HINMINE leverages weighting schemes used in text analysis for word weighting in BoW vector construction (introduced in Sect. 3.2.1) to weighting nodes in network analysis.

Consequently, the underlying idea of HINMINE is to treat the paths between network nodes as if they were words in text documents. For this aim, it forms BoW representations by treating paths (with one intermediate node) between pairs of nodes of interest as individual words. It samples the paths and counts their occurrences. The counts are weighted using appropriate weighting heuristics (explained below), ensuring that more specific paths are assigned higher weights. In HINMINE, Eq. 5.3 is therefore adapted and extended in a similar way as the term frequency (TF) weights are adapted in text mining. As a result, we replace count

1 with weight $w(t)$, where w is a weighting function which e.g., penalizes terms t appearing in many documents (e.g., the TF-IDF weighting function presented in Eq. 3.1 in Sect. 3.2.2) or which e.g., penalizes terms t appearing in documents from different classes (such as χ^2 mentioned in Sect. 3.2.2).

Based on this reasoning, in the calculation of $w(v, u)$ in the network setting, we replace the count 1 in Eq. 5.3 with a function $w(m)$, which penalizes intermediate nodes m that link to many base nodes or those that link base nodes from different classes. The result is the following formula for computing the weight of a link between nodes v and u:

$$w(v, u) = \sum_{\substack{m \in M \\ m \text{ is linked to } v \text{ and } u}} w(m). \tag{5.4}$$

By using different weighting functions w in Eq. 5.4, more informative weights can be used in the network decomposition step, described in Sect. 5.3.1.1. Analogous to the TF-IDF function, weighting functions w should decrease the importance of links to/from highly connected intermediate nodes. To this end, the term weighting schemes (described in Sect. 3.2.2) can be adapted to set weights to intermediate nodes in heterogeneous networks (such as authors in our example). This can be done in a way that the weight of a link between two nodes is the sum of weights of all the intermediate nodes they share, as defined in Eq. 5.4. Moreover, if we construct a homogeneous network in which nodes are connected if they share a connection to a node of type T in the original heterogeneous network, then the weight of the link between nodes v and u is equal to:

$$w(v, u) = \sum_{\substack{m \in T: \\ (m,v) \in E \\ (m,u) \in E}} w(m), \tag{5.5}$$

where $w(m)$ is the weight assigned to the intermediate node m.

The value of $w(m)$ can be calculated in several ways. Table 5.1 shows the intermediate node weighting heuristics corresponding to term weightings used in text mining and information retrieval, which were briefly outlined in Sect. 3.2.2. Table 5.1, which includes textual descriptions of the original term weightings and their network analysis variants developed in HINMINE, includes the formulas for TF, TF-IDF and χ^2 network analysis adaptations. The interested reader can find the formulas for the IG, GR, Delta-IDF, RF, and Okapi BM25 inspired node weighting schemes in Kralj et al. (2018). The TF weight, where all the intermediate nodes (i.e. authors) are weighed equally, was effectively used in the TEHmINe methodology.

The empirical results of HINMINE (Kralj et al. 2018) show that the choice of heuristics impacts the performance of classifiers that use the TEHmINe propositionalization approach (Grčar et al. 2013). However, the choice of the best weighting heuristic depends on the structure of the network. We demonstrate the computation of the χ^2 heuristics on an illustrative example network shown in Fig. 5.9.

Table 5.1 Three selected heuristics for weighting intermediate nodes in the decomposition step of constructing homogeneous networks (TF, TF-IDF and χ^2), and the comparison of these HINMINE network analysis heuristics with their original text mining counterparts. The notation is as follows: B is the set of all nodes of the base node type, E is the set of all edges in the heterogeneous network, m is a node, c is a class label, C is a set of all class labels, and P denotes probability

Scheme	Formula
TF	1
TF-IDF	$\log\left(\dfrac{\|B\|}{\|\{b \in B : (b,m) \in E\}\|}\right)$
χ^2	$\displaystyle\sum_{c \in C} \dfrac{(P(m \wedge c) \cdot P(\neg m \wedge \neg c) - P(m, \neg c) \cdot P(\neg m, c))^2}{P(m) \cdot P(c) \cdot P(\neg m) \cdot P(\neg c)}$

Scheme	Text mining context	Network analysis context
TF	Counts each appearance of each word in a document equally	Counts each intermediate node connecting two base nodes equally
TF-IDF	Gives a greater weight to a word if it appears only in a small number of documents	Gives a greater weight to an intermediate node if it is connected to only a small number of base nodes
χ^2	Given a term, measures how dependent the class label of a document is to the appearance of the given word in the document	Given an intermediate node, measures how dependent the class label of a base node is to the existence of a link between the base node and the given intermediate node

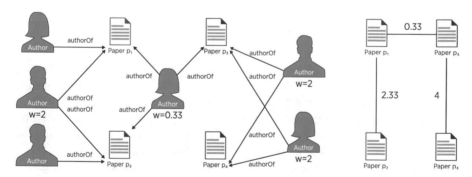

Fig. 5.9 The construction of a homogeneous network from the toy network on the left/hand side using the χ^2 heuristic. The blue color of nodes denotes that the paper belongs to class 1 (e.g., the first topic) and the red color denotes class 2

Example 5.1 (Heterogeneous Network Transformation) Consider the heterogeneous network shown on the left-hand side of Fig. 5.8, taking into account the class labels indicated on the left-hand side of Fig. 5.9. Using the χ^2 heuristic, this network is transformed into the network shown on the right-hand side of Fig. 5.9.

The weight of the central author m is calculated as the sum over both classes (the blue and the red class) of values of two expression:

$$\frac{(P(m \wedge c) \cdot P(\neg m \wedge \neg c) - P(m, \neg c) \cdot P(\neg m, c))^2}{P(m) \cdot P(c) \cdot P(\neg m) \cdot P(\neg c)} \tag{5.6}$$

- For the first (blue) class, we calculate the required values as $P(m \wedge c) = \frac{2}{4}$, $P(\neg m \wedge \neg c) = \frac{1}{4}$, $P(m, \neg c) = \frac{1}{4}$, $P(\neg m \wedge c) = 0$; $P(m) = \frac{3}{4}$, $P(\neg m) = \frac{1}{4}$, $P(c) = P(\neg c) = \frac{1}{2}$. For the blue class, Expression 5.6 is computed as follows:

$$\frac{(\frac{2}{4} \cdot \frac{1}{4} - \frac{1}{4} \cdot 0)^2}{\frac{3}{4} \cdot \frac{1}{4} \cdot \frac{1}{2} \cdot \frac{1}{2}} = \frac{1}{3}.$$

- For the second (red class) we calculate $P(m \wedge c) = P(\neg m \wedge \neg c) = P(m, \neg c) = P(\neg m \wedge c) = \frac{1}{4}$. Therefore , the value of Expression 5.6 is 0.

As a result, the total weight of author m is $\frac{1}{3} + 0 = \frac{1}{3}$ (i.e. 0.33). The weights of the other intermediate nodes (authors) are computed in the same way. In our example, none of the other authors wrote papers from both classes, so their weights are all equal to 2.

The homogeneous network on the right-hand-side of Fig. 5.9 is constructed by summing up the weights of all common authors for each pair of papers. We can see that the connection between papers p_1 and p_3 is weaker than the other connections because the joint author was assigned a smaller weight.

5.4   Ontology Transformations

This section aims to acquaint the reader with the representation learning approaches that successfully leverage domain-curated background knowledge. This knowledge is often in the form of ontologies that have to be transformed into a form suitable for machine learning algorithms. One of the areas where incorporation of ontologies into machine learning is actively researched is *semantic data mining* (SDM) (Žáková et al. 2006; Lawrynowicz and Potoniec 2011; Hilario et al. 2011; Podpečan et al. 2011; Vavpetič and Lavrač 2012). In this section, we first give a background on ontologies, and outline their use in SDM in Sect. 5.4.1. In Sect. 5.4.2, we present the NetSDM methodology for ontology reduction that allows efficient SDM even with very large ontologies and datasets by constraining the search only to the most relevant parts of ontologies. This approach to ontology reduction is relevant also for other tasks in heterogeneous information networks, including embedding of feature rich graphs with GCNs addressed in Sect. 5.2.3, and heterogeneous network propositionalization addressed in Sect. 5.3.1. Nevertheless, such combined approaches are out of the scope of the present monograph.

5.4.1 Ontologies and Semantic Data Mining

With the introduction of the semantic web (Berners-Lee et al. 2001), ontologies emerged as the standard data structure for working with background knowledge. Formally, ontologies are directed acyclic graphs, formed of concepts and their relations, encoded as *subject-predicate-object* (s, p, o) triplets. In computational biology, examples of structured domain knowledge include the Gene Ontology (GO) (Ashburner et al. 2000), the KEGG ontology (Ogata et al. 1999) and the Entrez gene–gene interaction data (Maglott et al. 2005). As an illustrative example, widely used in bioinformatics, take the well-known Gene Ontology (Ashburner et al. 2000), which contains tens of thousands of interconnected biological terms, corresponding to cellular and functional processes between individual genes, as well as groups of genes described by higher-level ontology terms. A small segment of the Gene Ontology is shown in Fig. 5.10.

The simplest ontologies are *taxonomies*, where nodes are hierarchically ordered from more general nodes at higher levels of the hierarchy to more specific concepts at lower levels of the hierarchy. Nodes are connected by *is-a* or *part-of* relationships between nodes. Ontologies can include other (non-hierarchical) types of relations and represent richer data structures than taxonomies. In both ontologies and taxonomies, bottom level nodes describe individual instances (e.g., individual genes in the GO), while higher-level nodes describe concepts or groups of instances (e.g., groups of genes in the GO).

To find patterns in data annotated with ontologies, *semantic data mining* (SDM) algorithms (Dou et al. 2015; Lavrač and Vavpetič 2015) take as input a set of class labeled instances and the background knowledge encoded in the form of ontologies. Frequently, the goal of SDM is to find descriptions of target class instances as a set of rules of the form *TargetClass ← Explanation*, where *Explanation* is a logical conjunction of terms from the ontology. SDM has its roots in symbolic rule learning, subgroup discovery, and enrichment analysis. Recent achievements in the area of deep learning (LeCun et al. 2015) have resulted in a series of novel, *semantics-aware* learning approaches to previously well-established problems, e.g., entity resolution, author profiling, and recommendation systems, as well as to heterogeneous network embedding (see Sect. 5.2).

The publicly available SDM toolkit (Vavpetič and Lavrač 2012) includes two semantic data mining systems: SDM-SEGS and SDM-Aleph. SDM-SEGS is an extension of a domain-specific algorithm SEGS (Trajkovski et al. 2008) that allows for semantic subgroup discovery in gene expression data. SDM-Aleph, which is built using the popular inductive logic programming system Aleph (Srinivasan 2007) can accept any number of ontologies expressed in the Web Ontology Language (OWL) and uses them as background knowledge in the learning process.

The Hedwig system (Vavpetič et al. 2013) takes the best from both SDM-SEGS and SDM-Aleph. It uses a search mechanism tailored to exploit the hierarchical nature of ontologies. Furthermore, Hedwig can take into account background knowledge in the form of Resource Description Framework (RDF) triplets. Hed-

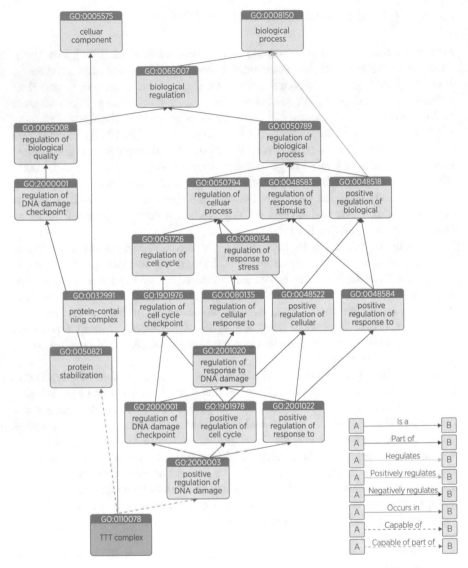

Fig. 5.10 A part of the Gene Ontology (Ashburner et al. 2000) related to cell regulation, with multiple types of relations (directed edges of different colors), demonstrating that ontologies are a specific form of heterogeneous information networks that include hierarchical relations

wig's output is a set of descriptive patterns in the form of rules, where rule conditions are conjunctions of domain ontology terms that explain a group of target class instances. Hedwig tries to discover explanations that best describe and cover as many target class instances and as few non-target instances as possible.

Table 5.2 An example output of the Hedwig SDM algorithm discovering subgroups. The Explanations represent subgroups of genes, differentially expressed in patients with breast cancer

Rank	Explanation
1	Chromosome ∧ cell type
2	Cellular macromolecule metabolic process ∧ Intracellular non-membrane-bounded organelle ∧ cell cycle
3	Cell division ∧ nucleus ∧ cell type
4	Regulation of mitotic cell cycle ∧ cytoskeletal part
5	Regulation of mitotic cell cycle ∧ microtubule cytoskeleton
6	Regulation of G2/M transition of mitotic cell cycle
7	Regulation of cell cycle process ∧ chromosomal part
8	Regulation of cell cycle process ∧ spindle
9	Enzyme binding ∧ regulation of cell cycle process ∧ Intracellular non-membrane-bounded organelle
10	ATP binding ∧ mitotic cell cycle ∧ nucleus

Example 5.2 (Rules Induced by Hedwig) To illustrate the rules induced by the Hedwig SDM algorithm (Vavpetič et al. 2013), take as an example a task of analyzing the differential expression of genes in breast cancer patients, originally addressed by Sotiriou et al. (2006). In this task, the *TargetClass* of the generated rules is the class of genes that are differently expressed in breast cancer patients when compared to the general population. The *Explanation* is a conjunction of Gene Ontology terms. It covers the set of genes annotated by the terms in the conjunction. For this problem, the Hedwig algorithm generated ten rules (subgroup descriptions), each describing a subgroup of differentially expressed genes. The resulting set of *Explanations* (rule conditions) is shown in Table 5.2. In the rule ranked first, the *Explanation* is a conjunction of biological concepts chromosome ∧ cell cycle, which covers all the genes annotated by both ontology terms. The fourth best-ranked rule explains that the mitotic cell cycle regulation and cytoskeletal formation explain breast cancer based on gene expression.

Let us explain the Hedwig system in some more detail. The semantic subgroup discovery task addressed by Hedwig takes three types of inputs: the training examples, the domain knowledge, and a mapping between the two. The *training examples* are expressed as RDF triplets in the form subject-predicate-object, e.g., geneX–suppresses–geneY. Dataset S is split into a set of positive examples S_+, i.e. 'interesting' target class instances (for example, genes enriched in a particular biological experiment), and a set of negative examples S_- of non-target class instances (e.g., non-enriched genes). The *domain knowledge* is composed of domain ontologies in the RDF form. The *annotations* are *object-to-ontology mappings* that associate RDF triplets with appropriate ontological concepts. An object x is *annotated by* ontology term o if the pair (x, o) appears in the mapping.

In addition, Aleph (A Learning Engine for Proposing Hypotheses) (Srinivasan 2007), which was briefly described in Sect. 4.4, is also incorporated into the SDM-toolkit (Vavpetič and Lavrač 2012), allowing Aleph (named SDM-Aleph) to use the same input data as Hedwig. A drawback of using SDM-Aleph is that it can reason only over one type of background knowledge relations, while the original Aleph algorithm can reason over any type of relations that allows Aleph to encode both the *is_a* and *part_of* relations, which are the most frequent relations in ontologies, as well as the relationships composed of these two relations.

5.4.2 NetSDM Ontology Reduction Methodology

SDM algorithms perform well on relatively small datasets. Even for small datasets, the algorithms search in a very large space of possible patterns. The more conjuncts we allow in rule conditions and the larger the background knowledge, the larger the search space. The NetSDM methodology (Kralj et al. 2019) constrains the search space to only the most relevant concepts. This is important as a mechanism of data and ontology reduction, which can be used in data preprocessing, before using an algorithm for network propositionalization or embeddings.

The overall NetSDM methodology, which is illustrated in Fig. 5.11, consists of four main steps:

1. Convert examples and background knowledge to network, where background knowledge terms represent nodes.
2. Estimate the significance of background knowledge terms using a node scoring function
3. Reduce the background knowledge by removing less significant terms, keeping only a proportion c of the top-ranking terms.
4. Apply a semantic data mining algorithm to the original dataset and the reduced background knowledge.

The NetSDM methodology improves the efficiency of SDM algorithms by removing the background knowledge terms that are not likely to appear in significant rules. The scoring function used in the ontology shrinkage should (i) evaluate the significance of terms based on the data, and (ii) be efficiently computed. In Sect. 5.4.2.2, we examine two possibilities for the scoring function, but first, we describe its intention and properties.

The input to the scoring function are the following data.

A set of instances S,   consisting of target (S_+) and non-target (S_-) instances.

Ontology O,   represented as a set of RDF triplets (*subject, predicate, object*). To simplify the notation, for each term t appearing as either a subject or object in a triplet, we write $t \in O$ to denote that t is a term of the ontology O.

Set of annotations $\mathscr{A}$, connecting instances in $S = S_+ \cup S_-$ with terms in O. The annotations are presented as triplets $(s, \textit{annotated-with}, t)$, where $s \in S$ and $t \in O$ denote that data instance s is annotated by an ontology term t.

The output of the scoring function is a vector, which—for each term in the background knowledge—contains a score estimating its significance for the dataset S. Using $\mathscr{T}(O)$ to denote the set of all terms appearing (either as subjects or objects) in ontology O, the scoring function maps the terms to non-negative real numbers

$$\mathrm{score}_{S,O,\mathscr{A}} : \mathscr{T}(O) \to [0, \infty).$$

The scores returned by the scoring function are used to shrink the background knowledge ontology O. A higher value assigned to term $t \in O$ means that term t is more likely to be used in rules describing the positive examples. This forms *relative ranks* of ontology terms:

$$\mathrm{RelativeRank}(t) = \frac{\mathrm{Rank}(t)}{|\mathscr{T}(O)|} = \frac{|\{t' \in \mathscr{T}(O) : \mathrm{score}(t') \geq \mathrm{score}(t)\}|}{|\mathscr{T}(O)|}.$$

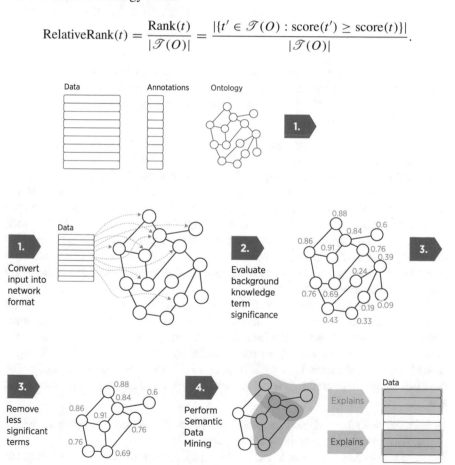

Fig. 5.11 Outline of the NetSDM methodology

Table 5.3 An illustrative
dataset consisting of 15
examples annotated by some
of the 8 possible annotations

Example	Class	Annotated by
1	+	A,B
2	–	A,B,C
3	+	B,C,D
4	+	B,C,E
5	+	B,C
6	–	C,D,E
7	+	D,E
8	+	D,E,F
9	–	E,F
10	–	E,F
11	–	F,G
12	–	F,G
13	–	G
14	+	G,H
15	+	G,H

The relative ranks are calculated as the ratio of terms in the ontology that score higher than t, divided by the number of all terms. For example, RelativeRank$(t) = 0.05$ means that only 5% of all terms in the ontology score higher than t using a given scoring function.

Example 5.3 (Illustrative Example) Consider the dataset described in Table 5.3. It consists of 15 instances, 7 of which belong to the target class (positive examples, marked with + in Table 5.3). Figure 5.12 illustrates the methodology. Using the simple ontology consisting of 8 base nodes and 7 higher-level terms, shown in the initial state (on the left-hand side of Fig. 5.12), the instances are annotated by one or more terms (Step 1). Using the Personalized PageRank scoring function (described in Sect. 5.4.2.2), high scores are assigned to the background knowledge terms that are strongly related to the positive examples (Step 2). We observe that the left-hand side terms have higher scores than the right-hand side terms as they annotate mostly the positive (target) examples.

In Step 3, the algorithm prunes the lower-ranked terms, leaving only the best 8 terms in the reduced hierarchy of ontology terms. As most of the top-scoring terms were in the left-hand part of the hierarchy, the reduced hierarchy mostly contains these terms. In the final Step 4, we use an SDM algorithm—in this case Hedwig— to find rules based on the instances from Table 5.3 and the reduced ontology. The best two rules discovered by Hedwig, which cover most of the positive examples, are:

Positive ← **LL**   (shown in green in Fig. 5.12), covering 3 positive and 1 negative instance (Coverage = 4, Precision = $\frac{3}{4}$).

Positive ← **LR**   (shown in red in Fig. 5.12), covering 5 positive and 2 negative instances (Coverage = 7, Precision = $\frac{5}{7}$).

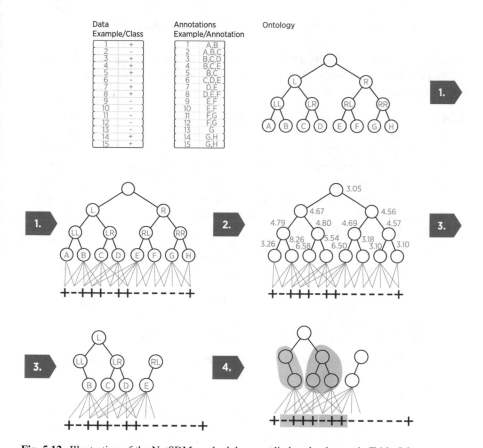

Fig. 5.12 Illustration of the NetSDM methodology applied to the dataset in Table 5.3

If we run a semantic data mining algorithm using the original (unpruned) ontology on this toy dataset, we get the same rules as on the reduced background knowledge. However, removing unimportant parts of large ontologies benefits the execution time and makes possible to analyze much larger datasets.

Below, we describe the first three steps of the NetSDM methodology (without the final learning step). In Sect. 5.4.2.1, we describe converting a background knowledge ontology into an information network, which we can then analyze using network analysis methods. In Sect. 5.4.2.2, we describe two methods for term importance estimation in information networks. In Sect. 5.4.2.3, we conclude with a discussion on the term removal step of the algorithm in which low-scoring terms are removed from the network.

5.4.2.1   Converting Ontology and Examples into Network Format

The first step of the NetSDM methodology is conversion of background knowledge ontologies into a network. We present two methods, a *direct* method and a *hypergraph* method.

Direct conversion.   Input ontologies for the NetSDM methodology are collections of triplets representing relations between background knowledge terms. As information networks are composed of nodes and the connections between them, a natural conversion from the background knowledge into an information network is to convert each relation between two ontology terms into an edge between the corresponding nodes in the information network. In the conversion, we merge the dataset and the ontology (background knowledge) into a single network $G_e = (V_e, E_e)$ where

- $V_e = \mathcal{T}(O) \cup S_+$ are vertices of the new network consisting of background knowledge terms $\mathcal{T}(O)$ and target class instances (positive examples) S_+;
- $E_e = \{(t_1, t_2)|\exists r : (t_1, r, t_2) \in O\} \cup \{(s, t)|(s, annotated\text{-}with, t) \in \mathcal{A}\}$ are edges of the new network consisting of edges that represent background knowledge relations, as well as edges that represent the *annotated-with* relations between data instances and background knowledge terms.

Hypergraph conversion.   Liu et al. (2013) proposed an alternative approach to converting an ontology into an information network format. They treat every relation in a semantic representation of data as a *(subject, predicate, object)* triplet. They construct a network (which they call a *hypergraph*) in which every triplet of the original ontology-based background knowledge (along with the background knowledge terms) forms an additional node in the network with three connections: one connection with the subject of the relation, one with the object of the relation, and one with the predicate of the relation. In this conversion, each predicate in the background knowledge is an additional node in the resulting network. This conversion results in a network $G_e = (V_e, E_e)$ where:

- $V_e = \mathcal{T}(O) \cup S_+ \cup \{n_r|r \in O \cup \mathcal{A}\} \cup \{p|\exists s, o : (s, p, o) \in O\} \cup \{annotated\text{-}with\}$ are vertices of the new network consisting of (i) background knowledge terms, (ii) target class instances, (iii) one new node n_r for each relation r either between background knowledge terms or linking background knowledge terms to data instances and (iv) one node for each predicate appearing in the ontology O, as well as (v) one node representing the *annotated-with* predicate;
- $E_e = \bigcup_{r=(s,p,o)\in O\cup\mathcal{A}} \{(s, n_r), (p, n_r), (o, n_r)\}$ are edges of the new network, each relation r inducing three connections to node n_r that represents this relation.

While hypergraph conversion results in a slightly larger information network, less information is lost in the conversion process compared to the direct conversion process. The hypergraph conversions create more connections between the ontology and the original network nodes. Therefore, more paths between nodes are annotated with similar parts of the ontology.

5.4.2.2 Term Significance Calculation

After converting the background knowledge and the input data into a network, we use a scoring function to evaluate the significance of background knowledge terms. Below we present two scoring functions, Personalized PageRank and node2vec. Term importance scoring functions $score_{S,O,\mathscr{A}}$ are constructed by taking into account dataset S, background knowledge terms $\mathscr{T}(O)$, and annotations $\mathscr{A}$.

P-PR-based score computation. While the basic PageRank algorithm (Page et al. 1999) evaluates the global importance of each node in a network, the Personalized PageRank evaluates the significance of a given node with respect to a given set of nodes (or a single node). This fits well with our demand that a scoring function has to consider the significance of a term based on the actual data. As each term $t \in \mathscr{T}(O)$ is a vertex in network G_e, we calculate its P-PR score as follows:

$$score_{P\text{-}PR}(t) = P\text{-}PR_{S_+}(t), \tag{5.7}$$

where the P-PR vector is calculated on the network G_e. A simple algebraic calculation (Grčar et al. 2013) shows that this value is equal to the average of all Personalized PageRank scores, where the starting set for the random walker is a node w from S_+:

$$P\text{-}PR_{S_+}(t) = \frac{1}{|S_+|} \sum_{w \in S_+} P\text{-}PR_{\{w\}}(t). \tag{5.8}$$

Following the definition of the Personalized PageRank score, the value of $score_{P\text{-}PR}$ is the stationary distribution of a random walker that starts its walk in one of the target data instances (elements of the set S_+) and follows the connections (either the is-annotated-by relation or a relation in the background knowledge). Another interpretation of $score_{P\text{-}PR}(t)$ is that it tells us how often we will reach node t in random walks, starting with positive (target class labeled) data instances. Note that the Personalized PageRank algorithm is defined on directed networks. This allows calculation of the Personalized PageRank by taking the direction of connections into account, but treating the network G_e as an undirected network is also possible. To calculate the P-PR vector on the undirected network, each edge between two nodes u and v is interpreted as a pair of directed edges, one going from u to v and another from v to u.

node2vec-based score computation. The second estimator of background knowledge terms significance is the node2vec algorithm (Grover and Leskovec 2016), described in Sect. 5.1.2. The default settings of $p = q = 1$ for the parameters of node2vec, mean that the generated random walks are balanced between the depth-first and breadth-first search of the network. As the default value, NetSDM uses the maximum length of random walks set to 15. The function node2vec (described in Sect. 5.1.2) calculates the feature matrix $f^* = \text{node2vec}(G_e)$, and each row of f^* represents a feature vector $f^*(u)$ for node u in G_e. The resulting feature vectors of nodes can be used to compute the similarity between any two nodes in the network. The approach uses the cosine similarity of feature vectors u and v:

$$\text{similarity}_{\text{node2vec}}(u, v) = \frac{f^*(u) \cdot f^*(v)}{|f^*(u)||f^*(v)|}.$$

NetSDM forms feature vectors for all nodes in the transformed graph (nodes representing both background knowledge terms and data instances). With these feature vectors one can compute the similarity between the background knowledge terms and the positive data instances (target class examples) using a formula inspired by Eq. 5.8. Recall that with the Personalized PageRank, we evaluate the score of each node as P-PR$_{S_+}(v)$, where S_+ is the set of all target class instances. As the value P-PR$_{\{w\}}(t)$ measures the similarity between w and t, we can replace the individual P-PR similarities in Eq. 5.8 with the individual node2vec similarities:

$$\text{score}_{\text{node2vec}}(t) = \frac{1}{|S_+|} \sum_{w \in S_+} \frac{f^*(w) \cdot f^*(t)}{|f^*(w)||f^*(t)|}. \qquad (5.9)$$

5.4.2.3   Network Node Removal

In the third step of the NetSDM methodology low-scored nodes are removed from the network. We present two variants of network node removal.

Naïve node removal.   The first (naïve) option is to take every low scoring node marked for removal and delete both the node and any edges. leading to or from it. This method is robust and can be applied to any background knowledge. However, if the background knowledge is heavily pruned, this method may cause the resulting network to decompose into several disjoint components. From the semantic point of view, the hypergraph representing the background knowledge has to contain relation nodes with *exactly* three neighbors (the subject, predicate, and object of the relation) if we are to map it back into a standard representation of the background knowledge. Thus, naïvely removing nodes from the hypergraph can result in a network no longer convertible back to the standard background knowledge representation.

Transitive node removal.   The relations encoded by the network edges are often transitive, i.e. $a\ R\ b\ \wedge\ b\ R\ c\ \implies\ a\ R\ c$. For example, the *is-a* and *part-of* relations are both transitive. Using the transitivity of relations, in the case that the middle node b is low scoring and shall be removed, we add the bridging connection $a\ R\ c$ to the graph and thereby assure its consistency.

With this idea, one can design an algorithm for removing low scoring nodes from information networks, obtained by the direct conversion of the background knowledge. The same algorithm can be used to remove low scoring nodes from hypergraphs constructed from the background knowledge. For this, we first have to convert hypergraphs into the same simple network form as produced by the direct conversion method.

Kralj et al. (2019) tested the NetSDM methodology on two real-world genetic datasets, the acute lymphoblastic leukemia and breast cancer. The acute lymphoblastic leukemia dataset, introduced by Chiaretti et al. (2004), contained a set of 1000 enriched genes (forming the target set of instances) from a set of 10,000 genes, annotated by concepts from the Gene Ontology (Ashburner et al. 2000), which was used as the background knowledge. In total, the dataset contained 167,339 annotations (connections between genes and Gene Ontology terms). The breast cancer dataset, introduced by Sotiriou et al. (2006), contains gene expression data of patients suffering from breast cancer. The set contains 990 interesting genes out of a total of 12,019 genes. The data instances are connected to the Gene Ontology terms by 195,124 annotations.

Using Hedwig as the semantic data mining algorithm for rule learning in the final step of the NetSDM methodology, the NetSDM discovered the same rules as the original Hedwig algorithm, being faster by a factor of almost 100. Using Aleph as the SDM algorithm on largely reduced background knowledge, the algorithm was still capable of discovering high-quality rules but being faster by a factor of 2. In general, the quality of the discovered rules decreased more quickly with Aleph compared to Hedwig when pruning larger and larger portions of the background knowledge.

5.5 Embedding Knowledge Graphs

In knowledge graphs (KG), edges correspond to relations between entities (nodes) and the graphs present subject-predicate-object *triplets*. The learning algorithms on KGs solve problems like triplet completion, relation extraction, and entity resolution. The KG embedding algorithms, briefly discussed below, are highly scalable and useful for large, semantics-rich graphs. For a detailed description and an extensive overview of the field, we refer the reader to Wang et al. (2017a), from where we summarize some of the key ideas underlying knowledge graph embedding.

In the below description of KG embedding algorithms, the subject-predicate-object triplet notation is replaced by the (h, r, t) triplet notation, where h is referred to as the *head* of the triplet, t as the *tail*, and r as the *relation* connecting the head and the tail. A schematic representation of triplet embedding is shown in Fig. 5.13. The embedding methods optimize the total plausibility of the input set of triplets, where plausibility of a single triplet is denoted with $f_r(h, t)$. This function shall be constructed in such a way that the triplets present in the graph are plausible.

Translational distance models.   These embedding approaches exploit distance-based scoring functions. They assess the plausibility of a fact with the distance between the two entities, usually after a translation carried out by the relation. One of the representative methods for this type of embedding is TransE (Bordes et al. 2013). TransE models relationships by interpreting them as translations operating on the low-dimensional embeddings of the entities. Relationships are represented as translations in the embedding space: if (h, r, t) relation holds, the embedding of the t entity should be close to the embedding of the h entity plus some vector that depends on the relationship r. The cost function being minimized can be stated as:

$$f_r(h, t) = ||h + r - t||^2.$$

For vectors h, r, and t in the obtained embedding, score $f_r(h, t)$ is close to zero if triplet (h, r, t) is present in the data.

Non-deterministic KG embeddings.   These KG embeddings take into account the uncertainty of observing a given triplet. A representative method for this type of embeddings is KG2E (He et al. 2015), which models the triplets with multivariate Gaussians. It models individual entities, as well as relations as vectors, drawn from multivariate Gaussians, assuming that h, r and t vectors are normally distributed, with mean vectors $\mu_h, \mu_r, \mu_t \in \mathbb{R}^d$ and covariance matrices $\Sigma_h, \Sigma_r, \Sigma_t \in \mathbb{R}^{d \times d}$, respectively. KG2E uses Kullback-Liebler divergence to directly compare the distributions as follows:

$$f_r(h, t) = \mathrm{KL}(\mathscr{N}(\mu_t - \mu_h, \Sigma_t + \Sigma_h), \mathscr{N}(\mu_r, \Sigma_r))$$
$$= \int \mathscr{N}_x(\mu_t - \mu_h, \Sigma_t + \Sigma_h) \ln \frac{\mathscr{N}_x(\mu_t - \mu_h, \Sigma_t + \Sigma_h)}{\mathscr{N}_x(\mu_r, \Sigma_r)} dx,$$

where $\mathscr{N}_x$ denotes the probability density function of the normal distribution.

Semantic matching models.   These embeddings exploit similarity-based scoring functions. They measure plausibility of facts by matching latent semantics of entities and relations embodied in their vector space representations. One of the

Fig. 5.13 Schematic representation of knowledge graph embedding. The head-relation-tail (h, r, t) triplets are used as inputs. Triplets are embedded in a common d-dimensional vector space

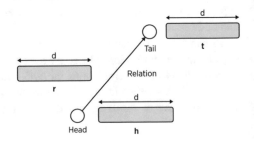

representative algorithms is RESCAL (Nickel et al. 2011) that optimizes the following expression:

$$f_r(h, t) = r^T \cdot M_r \cdot t,$$

where h and t are representations of entities, and $M_r \in \mathbb{R}^{d \times d}$ is a matrix associated with relations.

Matching using neural networks. Neural KG embeddings train neural networks and extract their weights as embeddings (similarly to neural text embeddings, presented in Sect. 3.3). One of the first approaches was Semantic Matching Energy (SME) (Bordes et al. 2014). This method first projects entities and their relations to their corresponding vector embeddings. The relation's representation is next combined with the relation's head and tail entities to obtain $g_1(h, r)$ and $g_2(t, r)$ entity-relation representations in the hidden layer. Finally, a dot product is used to score the triplet relation matching in the output layer:

$$f_r(h, t) = g_1(h, r)^T \cdot g_2(t, r).$$

The simplest version of SME defines the g_1 and g_2 as the weights of the hidden layer of a neural network:

$$g_1(h, r) = W_1^{(1)} \cdot h + W_1^{(2)} \cdot r + b_1$$

$$g_2(t, r) = W_2^{(1)} \cdot t + W_2^{(2)} \cdot r + b_2.$$

Here, $W_1^{(1)}$, $W_1^{(2)}$, $W_2^{(1)}$ and $W_2^{(2)}$ are $\mathbb{R}^{d \times d}$ dimensional weight matrices and b_1 and b_2 are bias vectors.

Other advances in embeddings of knowledge graphs also demonstrate interesting research directions. For example, hyperbolic geometry is used to better capture latent hierarchies, commonly present in real-world graphs (Nickel and Kiela 2017). KG embedding methods are increasingly tested on large, multi-topic data collections, for example, the Linked Data (LD) that standardize and fuse data from different resources. KG embeddings, such as RDF2vec (Ristoski and Paulheim 2016), exploit vast amounts of information in LD and transform it into a learning-suitable format. As knowledge graphs are not necessarily the only source of available information, algorithms exploit also other information, e.g., textual information available for each triplet (Wang et al. 2014). Other trends in knowledge graph embeddings explore how symbolic, logical structures could be used during embedding construction. Approaches like KALE (Guo et al. 2016) construct embeddings by taking into account logical rules (e.g., Horn clauses) related to the knowledge graph, thus increasing the quality of embeddings. Similar work was proposed by Rocktäschel et al. (2015), where pairs of embeddings were considered during optimization. The same research group showed how relations can be modeled without grounding the head and tail entities for simple implication-like clauses (Demeester et al. 2016). Wang et al. (2015) demonstrated that logical rules can aid in knowledge graph completion on large knowledge bases. They showed that the inclusion of rules can reduce the solution space and significantly improve the inference accuracy of embedding models.

5.6   Implementation and Reuse

This section demonstrates two popular network embedding methods: node2vec and metapath2vec. Node2vec works on homogeneous networks, while metapath2vec extends the scope to heterogeneous networks by introducing meta-paths.

5.6.1   Node2vec

Algorithm node2vec (Grover and Leskovec 2016), introduced in Sect. 5.1.2, is a popular method for embedding homogeneous networks based on the idea that nodes of a graph can be treated as words and random walks as sentences, thus extending the word2vec approach to network embedding. This idea was first implemented in the DeepWalk algorithm (Perozzi et al. 2014) and has since become a useful tool to convert networks into a tabular format where each row is an embedding of a node into a fixed-dimensional space. Node2vec is demonstrated on a publicly available network that contains dependencies among Python packages. A Jupyter notebook with code is available in the repository of this monograph: https://github.com/vpodpecan/representation_learning/blob/master/Chapter5/node2vec.ipynb.

5.6.2   Metapath2vec

Algorithm metapath2vec (Dong et al. 2017), introduced in Sect. 5.2.4, extends the idea of DeepWalk and node2vec to heterogeneous networks. It introduces meta-path based random walks to generate heterogeneous neighborhoods. In this way, semantic relationships can be incorporated into the skip-gram embeddings model. Metapath2vec is demonstrated on a publicly available heterogeneous DBLP citation network containing three node types: authors, papers, and venues. A Jupyter notebook with code is available in the repository of this monograph: https://github.com/vpodpecan/representation_learning/blob/master/Chapter5/metapath2vec.ipynb.

5.6.3 HINMINE

Algorithm HINMINE (Kralj et al. 2018), outlined in Sect. 5.3.2, can decompose a heterogeneous information network into homogeneous networks. This is accomplished using decomposition cycles and a term weighting scheme. Once the network is decomposed, any existing homogeneous network approach can be used, e.g., Personal PageRank. As a result, computed vector representations of nodes (embeddings) can be used in node classification tasks. A Jupyter notebook that applies HINMINE on the IMDB movie dataset and performs movie genre classification is available in the repository of this monograph: https://github.com/vpodpecan/representation_learning/blob/master/Chapter5/hinmine.ipynb

References

Monika Žáková, Filip Železný, Javier A. Sedano, Cyril Masia Tissot, Nada Lavrač, Petr Kremen, and Javier Molina. Relational data mining applied to virtual engineering of product designs. In *Proceedings of the 16th International Conference on Inductive Logic Programming (ILP'06)*, pages 439–453, 2006.

Michael Ashburner, Catherine A. Ball, Judith A. Blake, David Botstein, Heather Butler, J. Michael Cherry, Allan P. Davis, Kara Dolinski, Selina S. Dwight, Janan T. Eppig, Midori A. Harris, David P. Hill, Laurie Issel-Tarver, Andrew Kasarskis, Suzanna Lewis, John C. Matese, Joel E. Richardson, Martin Ringwald, Gerald M. Rubin, and Gavin Sherlock. Gene Ontology: Tool for the unification of biology. *Nature Genetics*, 25(1):25, 2000.

Tim Berners-Lee, James Hendler, and Ora Lassila. The semantic web. *Scientific American*, 284 (5):34–43, May 2001.

Antoine Bordes, Nicolas Usunier, Alberto Garcia-Duran, Jason Weston, and Oksana Yakhnenko. Translating embeddings for modeling multi-relational data. In *Advances in Neural Information Processing Systems*, pages 2787–2795, 2013.

Antoine Bordes, Xavier Glorot, Jason Weston, and Yoshua Bengio. A semantic matching energy function for learning with multi-relational data. *Machine Learning*, 94(2):233–259, 2014.

Joan Bruna, Wojciech Zaremba, Arthur Szlam, and Yann LeCun. Spectral networks and locally connected networks on graphs. In *International Conference on Learning Representations ICLR2014*, 2014.

Shiyu Chang, Wei Han, Jiliang Tang, Guo-Jun Qi, Charu C Aggarwal, and Thomas S Huang. Heterogeneous network embedding via deep architectures. In *Proceedings of the 21th ACM SIGKDD International Conference on Knowledge Discovery and Data Mining*, pages 119–128, 2015.

Fenxiao Chen, Yun-Cheng Wang, Bin Wang, and C-C Jay Kuo. Graph representation learning: A survey. *APSIPA Transactions on Signal and Information Processing*, 9, 2020.

Haochen Chen, Bryan Perozzi, Yifan Hu, and Steven Skiena. HARP: Hierarchical representation learning for networks. In *Proceedings of AAAI'2018*, 2018.

Sabina Chiaretti, Xiaochun Li, Robert Gentleman, Antonella Vitale, Marco Vignetti, Franco Mandelli, Jerome Ritz, and Robin Foa. Gene expression profile of adult T-cell acute lymphocytic leukemia identifies distinct subsets of patients with different response to therapy and survival. *Blood*, 103(7):2771–2778, 2004.

Michaël Defferrard, Xavier Bresson, and Pierre Vandergheynst. Convolutional neural networks on graphs with fast localized spectral filtering. In *Proceedings of the 30th International Conference on Neural Information Processing Systems*, pages 3844–3852, 2016.

Thomas Demeester, Tim Rocktäschel, and Sebastian Riedel. Lifted rule injection for relation embeddings. In *Proceedings of the 2016 Conference on Empirical Methods in Natural Language Processing*, pages 1389–1399, 2016.

Yuxiao Dong, Nitesh V Chawla, and Ananthram Swami. metapath2vec: Scalable representation learning for heterogeneous networks. In *Proceedings of the 23rd ACM SIGKDD International Conference on Knowledge Discovery and Data Mining*, pages 135–144, 2017.

Yuxiao Dong, Ziniu Hu, Kuansan Wang, Yizhou Sun, and Jie Tang. Heterogeneous network representation learning. In *Proceedings of the 2020 International Joint Conferences on Artifical Intelligence, IJCAI.*, pages 4861–4867, 2020.

Dejing Dou, Hao Wang, and Haishan Liu. Semantic data mining: A survey of ontology-based approaches. In *Proceedings of the 2015 IEEE International Conference on Semantic Computing (ICSC)*, pages 244–251, 2015.

Jordi Duch and Alex Arenas. Community detection in complex networks using extremal optimization. *Physical Review E*, 72(2):027104, 2005.

Tao-Yang Fu, Wang-Chien Lee, and Zhen Lei. Hin2vec: Explore meta-paths in heterogeneous information networks for representation learning. In Ee-Peng Lim et al., editor, *Proceedings of the 2017 ACM on Conference on Information and Knowledge Management, CIKM 2017, Singapore, November 06–10, 2017*, pages 1797–1806. ACM, 2017.

Palash Goyal and Emilio Ferrara. Graph embedding techniques, applications, and performance: A survey. *Knowledge-Based Systems*, 151:78–94, 2018.

Miha Grčar, Nejc Trdin, and Nada Lavrač. A methodology for mining document-enriched heterogeneous information networks. *The Computer Journal*, 56(3):321–335, 2013.

Aditya Grover and Jure Leskovec. node2vec: Scalable feature learning for networks. In *Proceedings of the 22nd ACM SIGKDD International Conference on Knowledge Discovery and Data Mining*, pages 855–864, 2016.

Shu Guo, Quan Wang, Lihong Wang, Bin Wang, and Li Guo. Jointly embedding knowledge graphs and logical rules. In *Proceedings of the 2016 Conference on Empirical Methods in Natural Language Processing*, pages 192–202, 2016.

William L. Hamilton, Rex Ying, and Jure Leskovec. Inductive representation learning on large graphs. In *Proceedings of Neural Information Processing Systems, NIPS*, 2017.

Shizhu He, Kang Liu, Guoliang Ji, and Jun Zhao. Learning to represent knowledge graphs with Gaussian embedding. In *Proceedings of the 24th ACM International on Conference on Information and Knowledge Management*, pages 623–632, 2015.

Melanie Hilario, Phong Nguyen, Huyen Do, Adam Woznica, and Alexandros Kalousis. Ontology-based meta-mining of knowledge discovery workflows. In *Meta-learning in computational intelligence*, pages 273–315. Springer, 2011.

Thomas N. Kipf and Max Welling. Variational graph auto-encoders. *NIPS Workshop on Bayesian Deep Learning*, 2016.

Thomas N. Kipf and Max Welling. Semi-supervised classification with graph convolutional networks. In *International Conference on Learning Representations (ICLR)*, 2017.

Jan Kralj, Marko Robnik-Šikonja, and Nada Lavrač. HINMINE: Heterogeneous information network mining with information retrieval heuristics. *Journal of Intelligent Information Systems*, 50(1):29–61, 2018.

Jan Kralj, Marko Robnik-Šikonja, and Nada Lavrač. NetSDM: Semantic data mining with network analysis. *Journal of Machine Learning Research*, 20(32):1–50, 2019.

Nada Lavrač and Anže Vavpetič. Relational and semantic data mining. In *Proceedings of the Thirteenth International Conference on Logic Programming and Nonmonotonic Reasoning*, pages 20–31, 2015.

Agnieszka Lawrynowicz and Jedrzej Potoniec. Fr-ONT: An algorithm for frequent concept mining with formal ontologies. In *Proceedings of 19th International Symposium on Methodologies for Intelligent Systems*, pages 428–437, 2011.

Yann LeCun, Yoshua Bengio, and Geoffrey Hinton. Deep learning. *Nature*, 521(7553):436, 2015.

Haishan Liu, Dejing Dou, Ruoming Jin, Paea LePendu, and Nigam Shah. Mining biomedical ontologies and data using RDF hypergraphs. In *Proceedings of the 12th International Conference on Machine Learning and Applications (ICMLA)*, volume 1, pages 141–146, 2013.

Donna Maglott, Jim Ostell, Kim D. Pruitt, and Tatiana Tatusova. Entrez Gene: Gene-centered information at NCBI. *Nucleic Acids Research*, 33:D54–D58, 2005.

Tomas Mikolov, Ilya Sutskever, Kai Chen, Greg S. Corrado, and Jeff Dean. Distributed representations of words and phrases and their compositionality. In *Advances in neural information processing systems*, pages 3111–3119, 2013.

Maximilian Nickel, Volker Tresp, and Hans-Peter Kriegel. A three-way model for collective learning on multi-relational data. In *Proceedings of International Conference on Machine Learning*, volume 11, pages 809–816, 2011.

Maximillian Nickel and Douwe Kiela. Poincaré embeddings for learning hierarchical representations. In *Advances in Neural Information Processing Systems*, pages 6338–6347, 2017.

Hiroyuki Ogata, Susumu Goto, Kazushige Sato, Wataru Fujibuchi, Hidemasa Bono, and Minoru Kanehisa. KEGG: Kyoto encyclopedia of genes and genomes. *Nucleic Acids Research*, 27(1): 29–34, 1999.

Lawrence Page, Sergey Brin, Rajeev Motwani, and Terry Winograd. The PageRank citation ranking: Bringing order to the web. Technical report, Stanford InfoLab, November 1999.

Bryan Perozzi, Rami Al-Rfou, and Steven Skiena. DeepWalk: Online learning of social representations. In *Proceedings of the 20th ACM SIGKDD International Conference on Knowledge Discovery and Data Mining*, pages 701–710, 2014.

Vid Podpečan, Nada Lavrač, Igor Mozetič, Petra Kralj Novak, Igor Trajkovski, Laura Langohr, Kimmo Kulovesi, Hannu Toivonen, Marko Petek, Helena Motaln, and Kristina Gruden. Seg-Mine workflows for semantic microarray data analysis in Orange4WS. *BMC Bioinformatics*, 12(1):416, 2011.

Jiezhong Qiu, Yuxiao Dong, Hao Ma, Jian Li, Kuansan Wang, and Jie Tang. Network embedding as matrix factorization: Unifying DeepWalk, LINE, PTE, and Node2Vec. In *Proceedings of the Eleventh ACM International Conference on Web Search and Data Mining*, WSDM '18, pages 459–467, 2018.

Alain Rakotomamonjy, Francis R. Bach, Stéphane Canu, and Yves Grandvalet. SimpleMKL. *Journal of Machine Learning Research*, 9:2491–2521, 2008.

Leonardo F. R. Ribeiro, Pedro H. P. Saverese, and Daniel R. Figueiredo. Struc2vec: Learning node representations from structural identity. In *Proceedings of the 23rd ACM SIGKDD International Conference on Knowledge Discovery and Data Mining*, KDD '17, pages 385–394, 2017.

Petar Ristoski and Heiko Paulheim. RDF2Vec: RDF graph embeddings for data mining. In Paul Groth, Elena Simperl, Alasdair Gray, Marta Sabou, Markus Krötzsch, Freddy Lecue, Fabian Flöck, and Yolanda Gil, editors, *The Semantic Web – ISWC 2016*, pages 498–514, 2016.

Tim Rocktäschel, Sameer Singh, and Sebastian Riedel. Injecting logical background knowledge into embeddings for relation extraction. In *Proceedings of the 2015 Conference of the North American Chapter of the Association for Computational Linguistics: Human Language Technologies*, pages 1119–1129, 2015.

Christos Sotiriou, Pratyaksha Wirapati, Sherene Loi, Adrian Harris, Steve Fox, Johanna Smeds, Hans Nordgren, Pierre Farmer, Viviane Praz, Benjamin Haibe-Kains, Christine Desmedt, Denis Larsimont, Fatima Cardoso, Hans Peterse, Dimitry Nuyten, Marc Buyse, Marc J. Van de Vijver, Jonas Bergh, Martine Piccart, and Mauro Delorenzi. Gene expression profiling in breast cancer: Understanding the molecular basis of histologic grade to improve prognosis. *Journal of the National Cancer Institute*, 98(4):262–272, 2006.

Ashwin Srinivasan. *The Aleph Manual*. University of Oxford, 2007. Online. Accessed 26 October 2020. URL: https://www.cs.ox.ac.uk/activities/programinduction/Aleph/.

Rainer Storn and Kenneth Price. Differential evolution: A simple and efficient heuristic for global optimization over continuous spaces. *Journal of Global Optimization*, 11(4):341–359, 1997.

Yizhou Sun and Jiawei Han. *Mining Heterogeneous Information Networks: Principles and Methodologies*. Morgan & Claypool Publishers, 2012.

Jian Tang, Meng Qu, and Qiaozhu Mei.   PTE: Predictive text embedding through large-scale heterogeneous text networks.   In *Proceedings of the 21th ACM SIGKDD International Conference on Knowledge Discovery and Data Mining*, pages 1165–1174, 2015a.

Jian Tang, Meng Qu, Mingzhe Wang, Ming Zhang, Jun Yan, and Qiaozhu Mei.   LINE: Large-scale information network embedding. In *Proceedings of the 24th International Conference on World Wide Web*, pages 1067–1077, 2015b.

Igor Trajkovski, Nada Lavrač, and Jakub Tolar. SEGS: Search for enriched gene sets in microarray data. *Journal of Biomedical Informatics*, 41(4):588–601, 2008.

Anže Vavpetič and Nada Lavrač.   Semantic subgroup discovery systems and workflows in the SDM-Toolkit. *The Computer Journal*, 56(3):304–320, 2012.

Anže Vavpetič, Petra Kralj Novak, Miha Grčar, Igor Mozetič, and Nada Lavrač.   Semantic data mining of financial news articles. In *Proceedings of Sixteenth International Conference on Discovery Science (DS 2013)*, pages 294–307, 2013.

Marinka Žitnik and Jure Leskovec.   Predicting multicellular function through multi-layer tissue networks. *Bioinformatics*, 33(14):i190–i198, 2017.

Quan Wang, Bin Wang, and Li Guo. Knowledge base completion using embeddings and rules. In *Proceedings of the 24th International Joint Conference on Artificial Intelligence*, pages 1859–1865, 2015.

Quan Wang, Zhendong Mao, Bin Wang, and Li Guo.   Knowledge graph embedding: A survey of approaches and applications. *IEEE Transactions on Knowledge and Data Engineering*, 29(12): 2724–2743, 2017a.

Xiao Wang, Peng Cui, Jing Wang, Jian Pei, Wenwu Zhu, and Shiqiang Yang.   Community preserving network embedding. In *Proceedings of the AAAI*, pages 203–209, 2017b.

Yaojing Wang, Yuan Yao, Hanghang Tong, Feng Xu, and Jian Lu.   A brief review of network embedding. *Big Data Mining and Analytics*, 2(1):35–47, 2020.

Zhen Wang, Jianwen Zhang, Jianlin Feng, and Zheng Chen.   Knowledge graph and text jointly embedding. In *Proceedings of the 2014 Conference on Empirical Methods in Natural Language Processing (EMNLP)*, pages 1591–1601, 2014.

Carl Yang, Yuxin Xiao, Yu Zhang, Yizhou Sun, and Jiawei Han.   Heterogeneous network representation learning: Survey, benchmark, evaluation, and beyond. *arXiv*, abs/2004.00216, 2020.

Siyi Zhu, Jiaxin Bing, Xiaoping Min, Chen Lin, and Xiangxiang Zeng. Prediction of drug–gene interaction by using metapath2vec. *Frontiers in Genetics*, 9, 2018.

Chapter 6
Unified Representation Learning Approaches

Throughout this monograph, different representation learning techniques have demonstrated that propositionalization and embeddings represent a multifaceted approach to symbolic or numeric feature construction, respectively. At the core of this similarity between different approaches is their common but *implicit* use of different similarity functions. In this chapter, we take a step forward by *explicitly* using similarities between entities to construct the embeddings. We start this chapter with Sect. 6.1, which presents entity embeddings, a general methodology capable of supervised and unsupervised embeddings of different entities, including texts and knowledge graphs. Next, two unified approaches to transforming relational data, PropStar and PropDRM, are presented in Sect. 6.2. These two methods combine propositionalization and embeddings, benefiting from both by capturing relational information through propositionalization and then applying deep neural networks to obtain dense embeddings. The chapter concludes by presenting selected methods implemented in Jupyter Python notebooks in Sect. 6.3.

6.1 Entity Embeddings with StarSpace

A general approach to harness embeddings is to use any similarity function between entities to form a prediction task for a neural network. Below we describe a successful *entity embedding* approach StarSpace (Wu et al. 2018). As this approach assumes discrete features from a fixed dictionary, it is particularly appealing to relational learning and inductive logic programming.

The idea of StarSpace is to form a prediction task, where a neural network is trained to predict the similarity between two related entities (e.g., an entity and its label or some other entity). The resulting neural network can be used for several purposes: directly in classification, to rank instances by their similarity, or using weights of the trained network as pretrained embeddings.

© Springer Nature Switzerland AG 2021
N. Lavrač et al., *Representation Learning*,
https://doi.org/10.1007/978-3-030-68817-2_6

In StarSpace, each entity has to be described by a set of discrete features from a fixed-length dictionary and forms a so-called *bag-of-features* (BoF). This representation is general enough for many different data modalities. For texts, documents or sentences can be described by bags-of-words or bags-of-n-grams; users can be described with bags of documents, movies, or items they like; relations and links in graphs can be described by semantic triplets. During training, entities of different kinds are embedded *in the same* latent space, suitable for various downstream learning tasks, e.g., a user can be compared with the recommended items. Note that entities can be embedded along with target classes, resulting in *supervised embedding learning*. This type of representation learning is the key element of the PropStar algorithm, presented in Sect. 6.2.1.

The StarSpace approach trains a neural network model to predict which pairs of entities are similar and which are dissimilar. Two kinds of training instances are formed, positive $(a, b) \in E^+$, which are task dependent and contain correct relations between entities (e.g., document a with its correct label b), and negative instances $(a, b_1^-), \ldots, (a, b_k^-) \in E_a^-$. For each entity a (e.g., a document), appearing in the set of positive instances E^+, negative instances are formed using k-negative sampling from labels $\{b_i^-\}_{i=1}^k$ as in word2vec (Mikolov et al. 2013). In each batch, a neural network aims to minimize the loss function L, defined as follows:

$$L = \sum_{(a,b) \in E^+} \left(\text{Loss}(\text{sim}(a, b)) - \frac{1}{k} \sum_{\substack{i=1 \\ (a,b_i^-) \in E_a^-}}^{k} \text{Loss}(\text{sim}(a, b_i^-)) \right),$$

where the function sim represents the similarity between the vector representations of the two entities. In the above expression, the function shall return large values for positive instances and values close to 0 for negative ones; often the dot product of the cosine similarity is used. For each batch update in neural network training, k negative examples (k being a parameter of the method) are formed by randomly sampling labels b_i^- from the set of entities that can appear as b. For example, in the document classification task, document a has its correct label b, while k negative instances have their labels b_i^- sampled from the set of all possible labels. The trained network predicts the similarity between the a and b. Within one batch, the instance loss function Loss sums the losses of the positive instance (a, b) and the average loss of the k negative instances (a, b_i^-), $i \in 1, \ldots, k$. To assess the loss, StarSpace uses margin ranking loss function Loss $= \max\{0, m - \text{sim}(a, b')\}$, where m is the margin parameter, i.e. the similarity threshold, and b' is any label.

The trained network can be used for several purposes. To classify a new instance a, one iterates over all possible labels b' and chooses $\arg\max_{b'} \text{sim}(a, b')$ as the prediction. For ranking, entities can be sorted by their predicted similarity score. The embedding vectors can also be extracted and used for other downstream tasks. Wu et al. (2018) recommended that the similarity function sim is shaped in such a way that it will directly fit the intended application so that training will be more effective.

A few examples of tasks successfully tackled with the StarSpace transformation approach are described below.

Multiclass text classification.   In this setting, the positive instances (a, b) are taken from the training set of documents E^+, represented with bags-of-words, and their labels b. For negative instances, entities b_i^- are sampled from the set of possible labels.

Recommender systems.   In this setting, users are described by a bag of items they liked (or bought). The positive instances use a single user identification as a and one of the items that the user liked as b. Negative instances take b_i^- from the set of possible items. Alternatively, to work for new users, the a part of user representation is a bag-of-items vector composed of all the items that the user liked, except one, which is used as b.

Link prediction.   For link prediction, the concepts in a graph are represented as *head-relation-tail* (h, r, t) triplets, e.g., gene-generates-protein. A positive instance a consists either of h and r, while b consists of t; alternatively, a consists of h, and b consists of r and t. Negative instances b_i^- are sampled from the set of possible concepts. The trained network can then predict links, e.g., gene-generates-X.

Sentence embedding.   In an unsupervised fashion of sentence embedding, a collection of documents containing sentences is turned into a training set. For positive instances, a and b are sentences from the same document (or are close together in a document), while for negative instances, sentences b_i^- are coming from different documents. This task definition tries to capture the semantic similarity between sentences in a document.

In the next section, we present transformations that combine propositionalization and embeddings. Note that the PropStar algorithm, described in Sect. 6.2.1, uses StarSpace similarly to the first case mentioned above (multiclass text classification).

6.2 Unified Approaches for Relational Data

The unifying aspects of propositionalization and embeddings introduced throughout this monograph, which will be summarized in Chap. 7, can be used as a basis for a unified methodology that combines propositionalization and embeddings and benefits from the advantages of both. While propositionalization successfully captures relational information through complex relational feature construction, it results in a sparse symbolic feature vector representation, which may not be optimal as input to contemporary learners. This weakness of propositionalization can be successfully overcome by embedding the constructed sparse feature vectors into a lower-dimensional numeric vector space, resulting in a dense numeric feature vector representation appropriate for deep learning.

Based on the work by Lavrač et al. (2020), this chapter presents two variants of these transformations, PropStar and PropDRM, both combining propositionalization and embeddings. Both approaches start from a sparse data representation of relational data, i.e. from sparse matrices returned by the wordification algorithm for relational data propositionalization (Perovšek et al. 2015), which is outlined in Sect. 4.5. In wordification, each instance is described by a bag (a multiset allowing multiple appearances of its elements) of features. The constructed features have the form *TableName_AttributeName_Value*. Wordification treats these simple interpretable features as *words* in the transformed bag-of-words data representation. In the two algorithms, PropStar and PropDRM, the specific implementation of the wordification approach to relational data propositionalization, presented in Sect. 4.5.3, is used.

The feature-based data transformation approach PropStar is presented in Sect. 6.2.1, while the instance-based data transformation approach PropDRM is presented in Sect. 6.2.2. The key difference between the two approaches is the type of embeddings: PropStar embeds features and class values in the same vector space, while PropDRM embeds instances (i.e. bags of constructed features).

6.2.1   PropStar: Feature-Based Relational Embeddings

The PropStar algorithm for relational data classification via feature embeddings is outlined below. The idea of PropStar is to generate embeddings that represent the *features* describing the dataset. Here, individual relational features, obtained as the result of propositionalization by wordification, presented in Sect. 4.5.3 are used by a supervised embeddings learner based on the SpaceStar approach (Wu et al. 2018) to obtain representations of features in the same latent space as the transformed instance labels. Opposed to the PropDRM approach, which will be presented in Sect. 6.2.2, where representations arc lcarncd for individual instances, in PropStar representations are learned separately for every relational feature returned from the wordification algorithm. In addition to embedding features, the labels (*true* and *false*) are also represented by vectors in the same dense space. The constructed embeddings leads to a simple classification of new instances. To classify a new instance, the embeddings of the set of its features (i.e. the features evaluated as *true* for the given instance) are averaged, and the result is compared to the embedding of the class labels. The nearest class label is chosen as the predicted class. This approach to classification is illustrated in Fig. 6.1.

In contrast to the instance-based embeddings used in PropDRM and discussed in Sect. 6.2.2, which relies on batches, the whole dataset is needed in PropStar to obtain representations for individual features. The StarSpace-based algorithms profits from operating on sparse bag-like inputs to avoid high spatial complexity. An example of feature-based embeddings are items recommended to users, where the representation of a given item is obtained by jointly optimizing the item's cooccurrence with other items, as well as other user's properties. In a relational setting using propositionalization by wordification, each instance is indeed described by a bag of features, which is already sparse and therefore very well-suited for the PropStar algorithm.

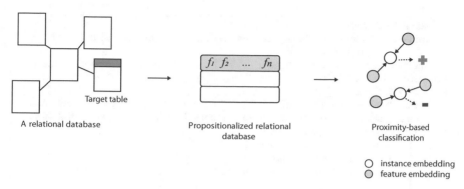

Fig. 6.1 Outline of the PropStar feature-based embedding methodology. For simplicity, the propositionalized database shows only two instances. Shaded circles correspond to embedded representations of features. Instance embeddings, represented with white circles, are averages of feature embeddings. The label vectors (positive and negative) are marked with the green and red cross. The closest label embedding to a given instance embedding determines the prediction. In the illustration above, one instance is classified as positive and the other as negative

The pseudocode of the PropStar algorithm is given in Fig. 6.2. The algorithm consists of two main steps. First, a relational database is transformed into sets of features describing individual instances (lines 1 to 20). The *wordify* method (line 8) constructs features of the form (table.name, column.name, value) and uses them to describe each individual instance (see Sect. 4.5.3 for a detailed formulation of this step). Second, sets of relational items (features) are used as input to the StarSpace entity embedding algorithm (line 21, described in Sect. 6.1), producing embeddings for each distinct relational item.

To apply the StarSpace algorithm, the problem is formulated as a multiclass classification. For each feature, the positive pair generator comes directly from a training set of labeled data specifying $(a, b) \in E^+$ pairs where a are multisets of features (sets of relational item acting as 'documents') and labels b are individual features (individual items). Negative entities b_i^- are sampled from the set of possible labels. Inputs can be described as multisets comprised of both relational items f_i, their conjuncts, and class labels c_i. Features of the form *TableName_AttributeName_Value* are constructed by the wordification algorithm, where these simple interpretable features are interpreted as *words* in the transformed bag-of-words data representation. In PropStar, these *words* represent individual *relational items*, using the (*table.name, column.name, value*) triplet notation format. For example, $\{f_1, f_2, f_6, f_6 \wedge f_2, c_1\}$ represents a simple input consisting of three features, a conjunct, and the target label c_1.

Note that we apply StarSpace in such a manner that the representations are learned for *individual relational items* (features). A representation matrix of dimension $\mathbb{R}^{|W| \times d}$ is produced as the final output ($|W|$ represents the number of unique relational items considered). Intuitively, the embedding construction can be understood as determining relational item locations in a latent space based on cooccurrence with other items present in all training instances. PropStar uses the

---

Data: A Relational database R, foreign key map m
Parameters   : Entity embedding parameter set $\mathscr{E}$, representation dimension d, target
                  table T, target attribute t
1 itemContainer ← empty bag of items ;   ▷ Begin wordification.
2 **foreach** *instance* $i \in T$ **do**
3 | relationalItems ← {} ;
4 | candidateKeys ← getForeignKeys(R,i) ;   ▷ Links to other tables.
5 | candidateTables ← getCandidateTables(candidateKeys, R) ;   ▷ Linked tables.
6 | **foreach** *table* $\in$ *candidateTables* **do**
7 | | **while** *not final number of items* **do**
8 | | | bagOfItems ← wordify(table(m(instance))) ;
9 | | | add bagOfItems to relationalItems ;   ▷ Store sampled items.
10 | | **end**
11 | **end**
12 | itemContainer[i].add(relationalItems) ;   ▷ Store relational items.
13 **end**
14 relevantFeatures ← frequencySelection(itemContainer.values, d) ;
15 symRep← [] ;   ▷ Sparse vector representations of instances.
16 **foreach** *instance* $i \in$ *targetTable* **do**
17 | instanceItems ← itemContainer[i] ;
18 | propRep ← RelationalFeatures(relevantFeatures, instanceItems) ;
19 | symRep.append(propRep) ;
20 **end**
21 *featureEmbeddings* ← StarSpace(symRep, $T[t], \mathscr{E}$) ; ▷ Input: a sparse matrix.
22 **return** *featureEmbeddings* ;

Fig. 6.2 The pseudocode of the PropStar algorithm

inner product similarity between a pair of vectors (representing relational items, i.e. features) e_1 and e_2 for the construction of embeddings: $\text{sim}(e_1, e_2) = e_1^T \cdot e_2$.

As the class labels are embedded in *the same* space as individual relational items, classification of novel bags of relational items is possible by direct comparison. Let m represent a novel instance to be classified, where m is represented as a multiset of relational items. StarSpace averages the representations of relational items (table.name, column.name, value), present in a given input instance (a bag). The representation is normalized (as during training) and compared to label embeddings in the common space. The representation of a relational bag e_m is computed as:

$$e_m = \frac{\underset{f_i \in m}{\oplus} e_{f_i}}{\sqrt{|m_{\text{unique}}|}},$$

which is a d-dimensional, real-valued vector. The $\oplus$ operator denotes an element-wise summation. The m_{unique} represents the set of all (unique) relational features considered in instance m. The resulting vector is compared to label embeddings in the same space. The label that is the most similar to e_m is the top-ranked prediction and gives the the following label assignment:

$$\text{label}(e_M) = \underset{c \in C}{\arg\max}[\text{sim}(\boldsymbol{e}_m, \boldsymbol{e}_c)].$$

The PropStar algorithm first samples the relational items with respect to the target table (lines 2–11 in Algorithm 6.2). Binary indicator function (relationalFeatures) stores the propositionalized representation of the target table (line 12). Zeros represent an absence of relational items and ones their presence. StarSpace embeds the table into a low-dimensional, real-valued embedding (line 19).

6.2.2 PropDRM: Instance-Based Relational Embeddings

The next unified approach to transforming relational databases, named PropDRM, is based on Deep Relational Machines (DRMs) (Dash et al. 2018), presented in Sect. 4.6, that use the output of the Aleph's feature construction approach, briefly mentioned in Sect. 4.4. PropDRM uses a variant of the DRM approach, which is capable of learning directly from large, sparse matrices that are returned by the wordification algorithm for propositionalizing relational databases (Perovšek et al. 2015), as outlined in Sect. 4.5 and implemented in Sect. 4.5.3 As mentioned before, in wordification each instance is described by a bag (a multiset that allows for multiple appearances of its elements) of features of the form *TableName_AttributeName_Value*. Wordification treats these simple interpretable features as *words* in the transformed bag-of-words data representation. In Prop-DRM, these *words* represent individual *relational items*, using the (*table.name, column.name, value*) triplet notation format.

In PropDRM, relational representations are obtained for individual instances, resulting in embeddings of instances (e.g., molecules, persons, or companies). Batches of instances are the input to a neural network, which performs the desired down-stream task, such as classification or regression. Schematically, the approach is illustrated in Fig. 6.3.

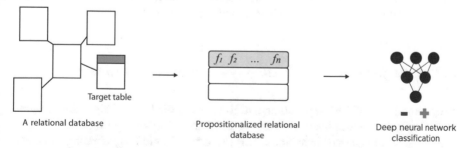

Fig. 6.3 Outline of the PropDRM instance-based embedding methodology, based on DRMs. Note that features in the propositionalized relational database can represent either single features f_i or conjuncts of features, e.g., $f_i \wedge f_j$ (wordification may construct both feature forms). For simplicity, the propositionalized database shows only two instances

Although propositionalization and subsequent learning are conceptually two distinct steps, they are not necessarily separated when implemented in practice. Neural networks are typically trained with small batches of input data; if propositionalization is capable of similar batch functionality, relational features can be generated in a lazy manner when needed by the neural network.

The main advantage of the PropDRM implementation is that it enables DRMs to operate on sparse matrices generated by the wordification algorithm. Let P represent a sparse item matrix returned by wordification. Wordification is unsupervised and thus does not include any information on instance labels. The neural network ω represents the mapping $\omega : P \rightarrow C$, where C is the set of classes.

PropDRM uses dense feed-forward neural networks, regularized using dropout (Srivastava et al. 2014), and ELU (Exponential Linear Unit) activation function on intermediary weights (for definition of ELU, see Eq. 2.4 in Sect. 2.1.7). The output weights are activated using the sigmoid function σ (see Eq. 2.1 in Sect. 2.1.7) to obtain binary predictions. For a given sparse input matrix P, a simple single hidden-layer neural network is defined as follows:

$$\omega = \sigma(W_1^T \cdot (\text{ELU}(\text{Drop}(W_0^T \cdot P + b_0))) + b_1),$$

where W_0 and W_1 are weight matrices, b_0 and b_1 are bias vectors of a given layer 0 or 1, and Drop represents dropout regularization.

The training procedure uses binary cross-entropy (CE) loss, which is defined for a given probabilistic classifier that returns a probability p_{ij} of an instance i belonging to class j, as follows:

$$\text{Loss}^{CE}(i) = \sum_{j \in C} y_{ij} \cdot \log p_{ij}.$$

where y_{ij} is a binary value (0 or 1) indicating whether class j is the correct class label assigned to instance i, and C is the set of all target classes. Each of the $|C|$ output neurons predicts a single probability p_{ij} for the given target class $j \in C$. If the neural networks are trained in small batches, the results of the Loss^{CE} function are averaged to obtain the overall loss of a given batch of instances.

6.2.3   Performance Evaluation of Relational Embeddings

PropStarand PropDRM were compared to propositionalization algorithms Aleph, RSD, RelF, and wordification, described in Chap. 4. They were empirically evaluated on several standard benchmark ILP (Inductive Logic Programming) datasets, published in previous work of Perovšek et al. (2015). It was shown that the two unified approaches, PropStar and PropDRM, perform competitively on most datasets, with their performance gains more pronounced on larger datasets (Lavrač et al. 2020). PropStar outperformed the other data transformation methods on three out of six tested datasets; therefore, PropStar can be considered a competitive candidate when addressing new relational problems. Most importantly, in terms of spatial requirements, compared to other propositionalization algorithms, only PropStar and PropDRM scale to large relational databases without specialized hardware.

6.3 Implementation and Reuse

This section demonstrates two general approaches: the entity embedding method StarSpace and the PropDRM method, which combines propositionalization and embeddings. StarSpace is demonstrated in a text classification scenario, while the well-known mutagenesis dataset (Srinivasan et al. 1994) is used to demonstrate PropDRM.

6.3.1 StarSpace

Entity embedding approach StarSpace (Wu et al. 2018), presented in Sect. 6.1, uses a similarity function between entities to construct a prediction task for neural embeddings. Objects of different types are projected into a common vector space, which allows their comparison. For example, in a recommendation task, a user representation is compared to an item representation to compute its desirability for the assessed user. In text classification, a document representation is compared to a label representation to select the best matching label. The Jupyter notebook, which reimplements the original StarSpace text classification example is available in the repository of this monograph: https://github.com/vpodpecan/representation_learning/blob/master/Chapter6/starspace.ipynb.

6.3.2 PropDRM

Algorithm PropDRM, presented in Sect. 6.2.2, is based on Deep Relational Machines (Dash et al. 2018) but employs wordification instead of Aleph for feature construction in the relational representation used by a neural network to perform a downstream task. The Jupyter notebook, demonstrating PropDRM on the mutagenesis dataset, is available in the repository of this monograph: https://github.com/vpodpecan/representation_learning/blob/master/Chapter6/propDRM.ipynb.

References

Tirtharaj Dash, Ashwin Srinivasan, Lovekesh Vig, Oghenejokpeme I Orhobor, and Ross D King. Large-scale assessment of deep relational machines. In *Proceedings of the International Conference on Inductive Logic Programming*, pages 22–37, 2018.

Nada Lavrač, Blaž Škrlj, and Marko Robnik-Šikonja. Propositionalization and embeddings: Two sides of the same coin. *Machine Learning*, 109:1465–1507, 2020.

Tomas Mikolov, Ilya Sutskever, Kai Chen, Greg S. Corrado, and Jeff Dean. Distributed representations of words and phrases and their compositionality. In *Advances in neural information processing systems*, pages 3111–3119, 2013.

Matic Perovšek, Anze Vavpetič, Janez Kranjc, Bojan Cestnik, and Nada Lavrač. Wordification: Propositionalization by unfolding relational data into bags of words. *Expert Systems with Applications*, 42(17–18):6442–6456, 2015.

Ashwin Srinivasan, Stephen H. Muggleton, Ross D. King, and Michael J. E. Sternberg. Mutagenesis: ILP experiments in a non-determinate biological domain. In *Proceedings of the 4th International Workshop on Inductive Logic Programming, volume 237 of GMD-Studien*, pages 217–232, 1994.

Nitish Srivastava, Geoffrey Hinton, Alex Krizhevsky, Ilya Sutskever, and Ruslan Salakhutdinov. Dropout: A simple way to prevent neural networks from overfitting. *Journal of Machine Learning Research*, 15(1):1929–1958, 2014.

Ledell Yu Wu, Adam Fisch, Sumit Chopra, Keith Adams, Antoine Bordes, and Jason Weston. StarSpace: Embed all the things! In *Proceedings of the 32nd AAAI Conference on Artificial Intelligence*, pages 5569–5577, 2018.

Chapter 7
Many Faces of Representation Learning

As this monograph demonstrates, albeit propositionalization and embeddings represent two different families of data transformations, they can be viewed as the two sides of the same coin. Their main unifying element is that they transform the input data into a tabular format and express the relations between objects in the original space as distances (and directions) in the target vector space. Our work indicates that both propositionalization and embeddings address transformations of entities with defined similarity functions as a multifaceted approach to feature construction. Based on the work by Lavrač et al. (2020), this chapter explores the similarities and differences of propositionalization and embeddings in terms of data representation, learning and use, in Sects. 7.1, 7.2 and 7.3, respectively. In Sect. 7.4 we summarize the strengths and limitations of propositionalization and embeddings, and conclude this monograph with some hints for further research.

7.1 Unifying Aspects in Terms of Data Representation

The unifying dimensions of propositionalization and embeddings in terms of data representation are summarized in Table 7.1.

Propositionalization. In propositionalization, the output of data transformation is a matrix of sparse binary vectors, where rows correspond to training instances, and columns correspond to symbolic features constructed by a particular propositionalization algorithm. These features are human interpretable, as they are either simple logical features (such as attribute values), conjunctions of such features, relations among simple features (such as, e.g., a test for the equality or inequality of values of two attributes of the same type), or relations among entities (such as links among nodes in a graph). Given that the number of constructed features is usually large, such transformation results in a sparse binary matrix with few non-zero elements.

© Springer Nature Switzerland AG 2021
N. Lavrač et al., *Representation Learning*,
https://doi.org/10.1007/978-3-030-68817-2_7

Table 7.1 Unifying and differentiating aspects of propositionalization and embeddings in terms of data representation

Representation	Propositionalization	Embeddings
Vector space	Symbolic	Numeric
Features/variables	Symbolic	Numeric
Feature values	Boolean	Numeric
Sparsity	Sparse	Dense
Space complexity	Space consuming	Mostly efficient
Interpretability	Interpretable	Non-interpretable

Embeddings.   Embeddings output is usually a dense matrix of a user-defined dimensionality, composed of vectors of numeric values, one for each entity of interest. For neural network based embeddings, vectors usually represent the activation of neural network nodes of one or more levels of a deep neural network. Given a relatively low dimensionality of these vectors (from 100 to 1000), such dense representations are efficient in terms of their space complexity. However, given that the constructed features (dimensions) are non-interpretable, separate explanation and visualization mechanisms are required.

7.2   Unifying Aspects in Terms of Learning

For both families of data transformation techniques, propositionalization and embeddings, the resulting tabular data representation is used as input to selected learning algorithms of the user's choice. The unifying dimensions of propositionalization and embeddings in terms of most frequently used learners are summarized in Table 7.2.

Propositionalization.   After propositionalization, any learner capable of processing tabular data described with symbolic features can be used. Typical learners include rule learning, decision tree learning, SVM, random forests or boosting for a supervised learning setting, as well as association rules, and symbolic clustering algorithms for a non-supervised learning setting. Learners usually use heuristic search, exception being, e.g., association rule learners using exhaustive search with constraints. Learners require some parameter tuning, but the number of (hyper)parameters for a typical learner is relatively low. Learning is typically performed on CPUs (Central Processing Units).

Embeddings.   After the embeddings transformation step, the resulting dense vector representation is best suited for learners, such as neural networks, and to a lesser degree for kernel methods or logistic regression. Deep neural networks use greedy search (typically stochastic gradient search) to find locally optimal solutions and are usually trained on GPUs (Graphic Processing Units) but can be used for prediction on both CPUs or GPUs. As a weakness, deep learning algorithms require substantial (hyper)parameter tuning.

Table 7.2 Unifying and differentiating aspects of propositionalization and embeddings in terms of learning context

Learning	Propositionalization	Embeddings
Capturing meaning	Via symbols	Via distances
Search strategy	Heuristic search	Gradient search
Typical algorithms	Symbolic	Deep neural networks
Tuning required	Some	Substantial
Hardware	Mostly CPU	Mostly GPU

7.3 Unifying Aspects in Terms of Use

The unifying dimensions of propositionalization and embeddings in terms of their use are summarized in Table 7.3.

Propositionalization. Propositionalization (Kramer et al. 2001) is one of the established methodologies used in relational learning (Džeroski and Lavrač 2001; De Raedt 2008) and inductive logic programming (Muggleton 1992; Lavrač and Džeroski 1994; De Raedt 2008). The propositionalization approach was also applied in semantic data mining where ontologies are used as background knowledge in relational learning (Podpečan et al. 2011; Lavrač et al. 2009; Vavpetič and Lavrač 2011). As for fusing heterogeneous data types, a propositionalization approach was proposed as a mechanism for heterogeneous data fusion (Grčar et al. 2013). Concerning the interpretability of results, propositionalization approaches are mostly used with symbolic learners whose results are interpretable as long as the models are small enough and given the interpretability of features used in the transformed data description.

Embeddings. Embedding technologies are mostly used in deep learning for various data formats, including tabular data, texts, images, and networks. As for data type fusion, several methods exploit heterogeneous networks for supervised embedding construction, e.g., PTE (Tang et al. 2015), metapath2vec (Zhu et al. 2018), and OhmNet (Žitnik and Leskovec 2017). For embedding-based methods, given the non-interpretable numeric features/dimensions, specific mechanisms need to be implemented to ensure results explanation. A well-known approach, which can be used in a post-processing phase of an arbitrary prediction model, is SHAP (Lundberg and Lee 2017), offering insights into instance-level predictions by assigning fair credit to individual features in prediction-explaining interactions. Hence, SHAP is commonly used to understand and debug black-box models.

Table 7.3 Unifying and differentiating aspects of propositionalization and embeddings in terms of use

Use	Propositionalization	Embeddings
Problems/context	Relational	Tabular, texts, graphs
Data type fusion	Enabled	Enabled
Interpretability	Directly interpretable	Special approaches

7.4 Summary and Conclusions

Let us summarize the presentation of propositionalization and embeddings by summarizing the strengths and limitations of the two data transformation approaches.

Strengths. The main strength of propositionalization is the interpretability of the constructed features and learned models. On the other hand, the main strength of embeddings is their compact vector space representation and high performance of classifiers learned from embeddings. Furthermore, both approaches are: (a) automated, (b) fast, (c) semantic similarity of instances is preserved in the transformed instance space, (d) transformed data can be used as input to standard propositional learners and contemporary deep learning approaches.
In addition to these characteristics, embeddings have other favorable properties: (a) embedded vectors representations allow for transfer learning, e.g., for cross-lingual applications in text mining or image classification from different types of images, (b) cover an extensive range of data types (text, relations, graphs, images, time series), and (c) have a vast community of developers and users, including industry.

Limitations. In terms of their limitations when used in a relational setting, both approaches to data transformation: (a) are limited to one-to-many relationships (cannot handle many-to-many relationships between the connected data tables), (b) cannot handle recursion, and (c) cannot be used for new predicate invention.
In addition to these characteristics, limitations of propositionalization include: (a) generated sparse vectors can be memory inefficient, (b) limited range of data types are handled (tables, relations, graphs), and (c) a small community of developers and users uses propositionalization tools in solving machine learning problems, mainly including ILP researchers.
Embeddings also have several limitations: (a) loss of explainability of features and consequently of the models trained on the embedded representations, (b) many user-defined hyper-parameters, (c) high memory consumption due to many weights in neural networks, and (d) requirement for specialized hardware (GPUs) for efficient training of embeddings, which may be out of reach for many researchers and end-users.

To conclude, we postulate that one of the main strategic challenges of machine learning is integrating knowledge and models across different domains and representations. While the research area of embeddings can unify different representations in a numeric space, symbolic learning is an essential ingredient for integrating different knowledge areas and data types. We perceive the unified approaches presented in Sect. 6.2 that combine advantages of propositionalization and embeddings in the same data fusion pipeline as a step in this direction.

One of the future research directions is how to better combine the explanation power of symbolic and neural transformation approaches. Indeed, the approximation power of deep neural network commonly used with embeddings comes at a cost of lesser interpretability. Compared to symbolic relational (or propositional) learners, one cannot understand the deep relational models' deductive process by inspecting the model. Currently, *post hoc* explanation methods, such as SHAP (Lundberg and Lee 2017), are used to better understand the parts of the feature space that are relevant for the neural network's individual predictions. Nevertheless, following the research by Lavrač et al. (2020), the state-of-the-art explanation tools based on the coalitional game theory, such as SHAP, should further be investigated to explain the learned models by explicitly capturing the interactions between the symbolic features.

Anther path worth investigating in the integration of symbolic and neural learning approaches is transfer learning, which can leverage many learning tasks and uses embeddings as input representations, and enrichment the embedding spaces with more information, either common sense knowledge or knowledge from specialized databases such as ontologies and knowledge graphs. In the perspective developed in this monograph, these developments will need to integrate many different relations into the embedded space. The ideas described in this monograph may provide useful insights how such integration can be addressed.

In summary, this monograph provides a comprehensive overview of transformation techniques for heterogeneous data propositionalization and embeddings in a wider machine learning and data mining context. We cover representation learning approaches for text, relational data, graphs, heterogeneous information networks, and other entities with defined similarity functions. We also cover more elaborate scenarios involving ontologies and knowledge graphs. The monograph aims to give intuitive presentations of the main ideas underlying individual representation learning approaches and their technical descriptions, supporting the explanations with reusable examples provided in the form of Jupyter Python notebooks.

References

Luc De Raedt. *Logical and Relational Learning*. Springer, 2008.

Sašo Džeroski and Nada Lavrač, editors. *Relational Data Mining*. Springer, Berlin, 2001.

Miha Grčar, Nejc Trdin, and Nada Lavrač. A methodology for mining document-enriched heterogeneous information networks. *The Computer Journal*, 56(3):321–335, 2013.

Stefan Kramer, Nada Lavrač, and Peter Flach. Propositionalization approaches to relational data mining. In Sašo Džeroski and Nada Lavrač, editors, *Relational Data Mining*, pages 262–291. Springer, 2001.

Nada Lavrač and Sašo Džeroski. *Inductive Logic Programming: Techniques and Applications*. Ellis Horwood, 1994.

Nada Lavrač, Petra Kralj Novak, Igor Mozetič, Vid Podpečan, Helena Motaln, Marko Petek, and Kristina Gruden. Semantic subgroup discovery: Using ontologies in microarray data analysis. In *Proceedings of the 31st Annual Intl. Conf. of the IEEE EMBS*, pages 5613–5616, 2009.

Nada Lavrač, Blaž Škrlj, and Marko Robnik-Šikonja. Propositionalization and embeddings: Two sides of the same coin. *Machine Learning*, 109:1465–1507, 2020.

Scott M Lundberg and Su-In Lee. A unified approach to interpreting model predictions. In *Advances in Neural Information Processing Systems*, pages 4765–4774, 2017.

Stephen H. Muggleton, editor. *Inductive Logic Programming*. Academic Press, London, 1992.

Vid Podpečan, Nada Lavrač, Igor Mozetič, Petra Kralj Novak, Igor Trajkovski, Laura Langohr, Kimmo Kulovesi, Hannu Toivonen, Marko Petek, Helena Motaln, and Kristina Gruden. Seg-Mine workflows for semantic microarray data analysis in Orange4WS. *BMC Bioinformatics*, 12:416, 2011.

Jian Tang, Meng Qu, and Qiaozhu Mei. PTE: Predictive text embedding through large-scale heterogeneous text networks. In *Proceedings of the 21th ACM SIGKDD International Conference on Knowledge Discovery and Data Mining*, pages 1165–1174, 2015.

Anže Vavpetič and Nada Lavrač. Semantic data mining system g-SEGS. In *Proceedings of the Workshop on Planning to Learn and Service-Oriented Knowledge Discovery (PlanSoKD-11), ECML PKDD Conference*, pages 17–29, 2011.

Marinka Žitnik and Jure Leskovec. Predicting multicellular function through multi-layer tissue networks. *Bioinformatics*, 33(14):i190–i198, 2017.

Siyi Zhu, Jiaxin Bing, Xiaoping Min, Chen Lin, and Xiangxiang Zeng. Prediction of drug–gene interaction by using metapath2vec. *Frontiers in Genetics*, 9, 2018.

Index

© Springer Nature Switzerland AG 2021
N. Lavrač et al., *Representation Learning*,
https://doi.org/10.1007/978-3-030-68817-2

Printed in the United States
by Baker & Taylor Publisher Services